Illu International Ham Radio Dictionary

Over 1500 radio terms and acronyms explained

First Edition

By Peter Parker VK3YE

VK3YE RADIO BOOKS

PETER PARKER VK3YE

ISBN-13: 9781793145505

CONTENTS

1 INTRODUCTION

Amateur radio's biggest strength is the diversity of activities within it. New activities and modes emerge each year and even experienced hams have trouble keeping up with terminology. The learning curve is even steeper for the person new or returning to radio.

I've prepared this dictionary to make your learning (or relearning) easier. From antennas to zener diodes, there are hundreds of definitions to make understanding easier. Many terms include illustrations or diagrams.

Dip randomly to expand your ham vocabulary. Or use it like a conventional dictionary. Keep it near the rig for those unfamiliar terms you hear on the air.

This is a first edition. Much listening, reading and research was done in the year it took to compile. For conciseness, only terms in common amateur radio usage are included. For the same reason wider physics, engineering or electronic definitions for some terms have been omitted.

This dictionary is for a general audience without a higher mathematics or engineering background. Consequently, I seek the indulgence of engineer readers who might find some entries lacking detail or over-simplified.

There will also inevitably be some missed or new definitions required as technology advances. I'd welcome suggestions for inclusion in any future edition through the *VK3YE Radio Books* Facebook page or email to vk3ye@qsl.net.

Peter Parker VK3YE

Melbourne Australia

December 2018

2 DEFINITIONS: A - Z

A

A: See *amp*.

A-index: A daily average measure of geomagnetic activity that gives an indication of disturbance to the ionosphere. Values can range from 0 (quiet) to 400 (very major storm). Ap index is a global value based on readings from multiple observatories. Also see *K index*.

A to D converter: An IC or stage that converts analogue signals to digital signals.

Absorption wavemeter: Test instrument used to indicate the relative strength and approximate frequency of a nearby RF signal and any harmonics. The stronger the incoming signal the more the meter needle moves. Similar to a field strength meter but contains an adjustable tuned circuit to make it frequency selective.

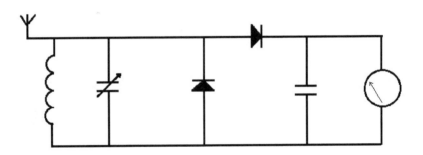

ABT: About. (CW abbreviation)

AC: Alternating current. Refers to a current that reverses in direction numerous times per second such as supplied from your power socket at home. Opposite to *DC* (direct current).

Access code: A keyed in code used to access certain facilities on an amateur radio repeater such as a link or autopatch.

Activator: A person who transmits from a sought-after location such as a SOTA summit, IOTA island or rare DX country to allow others to claim points for awards.

Active antenna: A small sized antenna containing an RF preamplifier. Often used for receiving when there is no room to erect a large antenna.

Active circuit: A circuit that requires a power source and active components to do its job. Examples in radio circuits include oscillators and amplifiers. Some filters, mixers and phase shift networks are also active circuits.

Active component: A component such as a transistor or IC able to generate or amplify a signal. Requires a power source to operate. Opposite to *passive component.*

Active filter: A filter containing active components such as transistors and ICs for the purpose of passing or rejecting audio frequencies at, above or below certain frequencies.

Actuator: The lever on a switch etc that when pressed causes contact to be made or broken.

Adapter: Item that converts one type of connection, voltage or information format to another.

Advanced licence: The highest licence level in Australia, conferring all amateur operating privileges.

Aerial: An *antenna*.

AF: Audio frequency. Refers to frequencies humans can hear such as in the 20 Hz to 20 kHz range.

AFSK: Audio frequency shift keying. A method of generating frequency shift keyed signals, as used for some text or digital-based modes. Can be produced by feeding an audio tone into an SSB transmitter and varying its pitch to produce the required frequency shift.

AGC: Automatic Gain Control. Circuitry in a receiver that automatically lowers its gain when set to a strong signal. This saves your ear drums when tuning from a weak signal to a strong signal. Better receivers have manually adjusted AGC controls with option including fast (acting), slow (acting) and off.

AGN: Again. (CW abbreviation)

AH: Ampere hour. Relates to the capacity of rechargeable batteries. For example, a 10 ampere hour battery should be able to deliver 1 amp of current for approximately 10 hours. Non-rechargeable batteries rarely have a marked AH rating.

Air dielectric: Refers to a tuning capacitor that has air between its metal plates. Opposite to solid dielectric.

Air wound: Refers to an inductor wound from sufficiently thick wire to be self-supporting and not require a former.

Aircraft scatter: Aircraft enhancement. Aircraft reflection). A technique that makes use of regularly scheduled aircraft to scatter radio signals and allow extended range VHF, UHF and microwave communication that would not otherwise be possible.

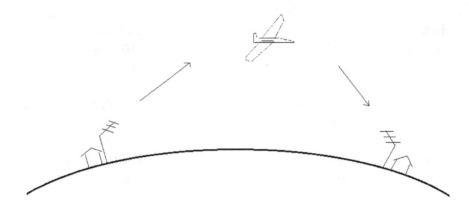

ALC: Automatic Level Control. Circuitry that prevents overload (and thus distortion) in an audio or radio frequency amplifier stage. A form of *AGC*.

ALF: Absorption limiting frequency. The lowest frequency that supports reliable communications via the ionosphere for a particular radio path.

Align: The act of making internal adjustments to a transmitter or receiver so it delivers its full performance. Alignment may involve adjusting internal oscillator frequencies, tuned circuits and signal levels. It is often done after repair or replacement of components. An attempt at alignment by a 'screwdriver expert' can easily result in misalignment. See *preset*.

Alkaline: Refers to a type of battery common for use with portable low current consumer items such as LED torches. Mostly non-rechargeable. Not recommended for applications with high peak current draw (such as radio transmitters) due to their high internal resistance.

All knobs to the right: A tendency of operators (especially 'screwdriver experts' or 'lids') to turn all controls to maximum in the hope that this will increase their signal. All that normally happens is their transmitter gets hot and signals distort.

All mode: Refers to a transceiver, transmitter or receiver that can handle all signal modes. Amongst amateurs CW, AM, FM and SSB capability is widely considered all mode since, at least on HF, other modes can be accommodated with a computer and suitable software.

AM: Amplitude Modulation. The original speech mode used for radio communication. Voice intelligence is conveyed by varying (modulating) the level (or amplitude) of the signal. AM signals comprises two amplitude modulated sidebands either side of a carrier. Often generated by using an audio amplifier (modulator) to vary the supply voltage of the transmitter's final amplifier stage. Largely replaced by SSB on HF in the 1960s and FM on VHF in the 1970s, AM retains a following amongst builders and restorers of broadcast and vintage equipment. A typical AM signal occupies a 6 to 8 kHz bandwidth.

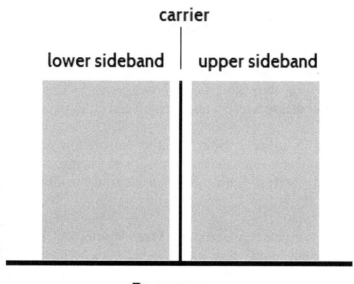

Amateur: An amateur radio operator or 'ham'. Radio amateurs have passed an examination in radio theory and regulations. They are authorised by government to transmit on allocated frequencies for self-training and intercommunications without pecuniary interest.

Amateurs Code: A code of conduct for radio amateurs written by US amateur Paul Segal in 1928. It emphasises values such as progressiveness, patriotism, balance and friendliness.

Ambient noise: Refers to externally generated RF noise audible on a receiver, such as generated by local electrical equipment. A city area will likely have higher noise than a rural area so will be an inferior receiving location. See *noise floor*.

Amp (abbreviation): See *amplifier*.

Amp (current): Ampere. Unit of electrical current. Refers to the maximum capacity of a power supply or the current draw of an electrical item.

Ammeter: Instrument that measures electrical current. Can be a separate instrument or a range on a multimeter.

Amplifier: One of the basic stages in radio circuits. The purpose of an amplifier is to make a weak signal stronger. This is required in transmitter and receivers at both audio and radio frequencies.

AMSAT: The amateur radio satellite corporation under whose auspices many amateur satellites are built and launched. Based in the United States with associated groups internationally. http://www.amsat.org

AMTOR: Amateur Teleprinting Over Radio. A 1970s print communications mode that grew out of radio teletype (RTTY) suited to poor HF conditions due to its error correction techniques. Largely replaced by more modern computer-based digital modes such as PSK31, FT8 and JS8 Call.

Analogue: Refers to transmissions or signals that are continuously variable in amplitude and/or frequency with an infinite range of values. Examples include the established voice modes such as AM, FM, and SSB. Opposite to *digital*.

Anderson Power Pole: A type of low voltage DC connector often used for transceiver power connections.

ANL: Automatic Noise Limiter (or Limiting). A circuit found in a receiver that limits noise such as impulse interference from vehicles.

Anode: Connection on a diode or electron tube that current flows in to. Opposite to cathode.

Antenna: One or more metal rods or wires used to collect or radiate radio signals in conjunction with a receiver or transmitter. Can be horizontal or vertical, large or small, omnidirectional or directional, monoband or broadband depending on need.

Antenna analyser: Piece of test equipment that allows testing of antennas for bandwidth and impedance. Has largely replaced dip oscillators and noise bridges for such measurements.

Antenna coupler: A capacitor and inductance network placed between the antenna and transceiver that transforms the impedance presented by the antenna to the 50 ohms required by the transceiver. This allows the transmitter to work into its specified impedance load and efficiently transfer signals. Normally adjusted with a VSWR meter between the coupler and transceiver.

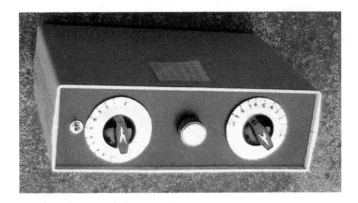

Antenna current: The current flow from a signal source (such as a transmitter) into the antenna. Measurable with an RF ammeter.

Antenna farm: Collection of masts and antennas such as found at an enthusiastic amateur's house.

Antenna modelling: The use of a computer program to simulate the operation of antennas, and assist in their design and performance optimisation. Popular antenna modelling programs include 4nec2, EZNEC, MMANA and MININEC.

Anti-static material: Foam-like material that protects sensitive transistors and ICs from damage due to static electricity. Reputable suppliers sell ICs inserted in such material while cheaper outlets sell them loose.

AOS: Acquisition of signal. Refers to when an orbiting satellite appears above the horizon and its signal becomes audible to stations on the ground at a particular location. Opposite to *LOS*.

Apogee: The highest point in the sky that a satellite reaches as observed from a point on earth.

APRS: Automatic Position Reporting System. A system that integrates GPS, computing and radio to allow vehicles and objects to be tracked.

AR (abbreviation): A common abbreviation for amateur radio.

AR (Morse signal): End of message in Morse code communication.

Arcing: Sparking between two conductors which, though separate, are sufficiently close for the insulating properties of the material between them (which can be air) to break down due to a high potential difference (i.e. voltage) between them. Observed across high voltage points of antennas such as magnetic loops where the variable capacitor has insufficient plate spacing for the voltage present.

ARDF: Amateur Radio Direction Finding. An amateur radio activity where participants equipped with receivers and directional antennas find a hidden transmitter ('the fox') either in vehicle or on foot.

Arduino: A popular and easily programmable open-source development board and software with many applications in computing, electronics and radio. https://www.arduino.cc/

ARES: Amateur Radio Emergency Service. An organisation of trained licenced amateurs who assist in public service and emergency communication. (USA) http://www.arrl.org/ares

ARISS: Amateur Radio on the International Space Station. An educational program sponsored by space agencies and amateur radio organisations that allows school students to speak to International Space Station crew. http://www.ariss.org

ARRL: American Radio Relay League. The national representative society for radio amateurs in the United States of America. Aims to represent the interests of amateurs to the FCC, provide member services, publishes QST magazine and supports the International Amateur Radio Union. http://www.arrl.org

ARRL DX Contest: One of the major international HF contests.

AS: Please wait. (CW abbreviation)

Att: Attenuator. A control on a radio receiver that allows you to lower its sensitivity. This is sometimes required if an incoming signal is so strong that it overloads the receiver and distorts reception. Attenuators are similar to RF gain controls except that the former is often one or two settings (e.g. minus 10 and 20 dB) while RF gain controls are continuously adjustable. See *RF Gain* and *RF Attenuator*.

Attack: The reaction time of an automatic gain control circuit. A fast attack time is desirable so that a strong signal does not cause a loud blast in the speaker.

Attenuate: To reduce the strength of a signal.

ATU: Antenna tuning unit. Transmatch. See *antenna coupler*.

ATV: Amateur television. Fast scan (i.e. moving picture) television signals sent by radio amateurs. Can be either digital or analogue.

Audio compressor: An electrical circuit incorporating an AGC circuit that boosts quiet signals and attenuates loud signal so that output is more constant. Used in microphone circuits to improve signal readability under weak signal conditions. Excessive compression can cause signals to be distorted or raise background noise. Also speech compressor.

Audio selectivity: Refers to the frequency response of audio amplifier stages. Whereas amplifiers for music require a wide frequency response, radio gear needs a narrower response centred on voice frequencies (see 'communications quality'). An audio filter can improve audio selectivity, particularly for narrow modes such as CW.

Auroral propagation: Extended distance VHF radio propagation that can occur in high latitude near-polar areas. Caused by ionisation resulting from auroral activity. Stations exploit these conditions by pointing their antennas towards the auroral region rather than towards each other.

Autopatch: A connection to a repeater that allows one to make telephone calls via amateur radio. Now largely superseded by widespread mobile phone ownership. (USA)

Autotransformer: A transformer with a single winding such as used to step AC voltage up or down. Cheaper and lighter than conventional two winding transformers with separate primary and secondary windings, autotransformers do not provide isolation between input and output.

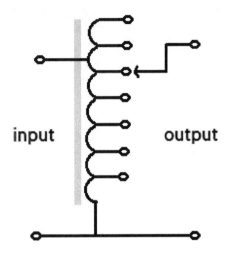

input   output

Auto tuner: An automatic antenna tuner. Often built into middle to high-end HF transceivers to compensate for minor impedance mismatches at the antenna socket. Auto tuners are also available as separate units. They can be mounted away from the transceiver but remotely controlled by it. Such placement at or near the antenna's feed point minimises feedline loss and allows a simple wire vertical or doublet to conveniently and efficiently operate on multiple bands.

Award: A certificate obtained for confirming contact with amateurs in a specific number of countries, islands, hill tops, national parks etc. Hundreds of awards, covering all facets of amateur radio, are available.

AWG: American Wire Gauge. A pre-metric method of measuring wire thickness. The larger the gauge number the finer the wire. SWG is the British measure which is slightly different.

AX.25: An amateur-developed protocol for packet radio data communication as widely used in the 1980s and 1990s.

Az-el rotator: Azimuth-elevation rotator. An advanced type of antenna rotator that allows both the azimuth and elevation of a rotary beam or dish antenna to be adjusted such as often required for satellite communication.

Azimuth: Relates to a compass direction. North is 0 (or 360 degrees), East is 90 degrees, South is 180 degrees and West is 270 degrees. Along with elevation this is used to describe the position of a satellite. Also see *elevation*.

Azimuthal map: A world map specially designed to show your location in the centre. Useful for calculating beam directions. It is possible to get a custom azimuthal map drawn online for your location - visit https://ns6t.net/azimuth/azimuth.html   Also see *great circle*.

Azimuthal Map
Center: 37°30'0"S 145°0'0"E
Courtesy of Tom (NS6T)

Map from http://ns6t.net

B

Backlash: Mechanical play in a radio tuning mechanism or dial that makes it harder to tune stations in. You may need to tune past a station then back several times to be spot on frequency.

Backlight: A light, often a globe or LED, that provides illumination for a radio dial or frequency display.

Backscatter: A radio propagation mode most experienced during high sunspot years capable of supporting medium distance communication on higher HF bands. This is a fairly high loss mode and signals received via backscatter are often fairly weak.

Backup power: A power source available when usual sources, such as your domestic mains supply, becomes unavailable due to natural disasters or system failures. Sources may include charged batteries, solar, wind or generators.

Backwave: A stray signal transmitted by a Morse code transmitter when the key is up and there is not supposed to be any transmission. May be caused by signals from the oscillator leaking through subsequent keyed stages.

Balanced antenna: A symmetrical antenna intended to be fed with a balanced feedline or via a balun to coaxial cable. Half wave dipoles and quad loops are common examples.

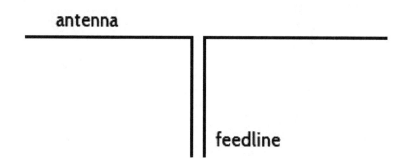

Balanced feedline: A form of feedline (such as open-wire) that is symmetrical and thus electrically balanced. Open wire features low loss, and in conjunction with a balanced antenna coupler, can allow a wire dipole to operate on multiple bands with minimal loss. Open wire is less robust than coaxial cable and should not be run alongside metal mast or guttering material.

Balanced mixer: A mixer stage, typically comprising RF transformers and diodes, which takes two input signals and produces at its output signals that are the sum and difference frequencies of the inputs. There exist single and double balanced variations. Single balanced mixers suppress one input at the output while double balanced mixers suppress both inputs, thus providing a cleaner output.

Balanced modulator: A balanced mixer circuit that generates a double sideband suppressed carrier voice signal on its output from audio and radio frequency carrier signals applied on its inputs. Used in DSB and SSB transmitters. Can be made from diodes, transistors or ICs.

Balloon amateur radio experiment: An activity where radio amateurs launch and track a transmitter in a helium balloon for educational or scientific experiments.

Balun: BALanced to UNbalanced. A broadband transformer that converts balanced feed to unbalanced feed, such as if you are connecting unbalanced coaxial feedline to a balanced antenna. May either be 1:1 (no impedance transformation) or various impedance transformation ratios such as 4:1 or 9:1. Can be wound on an air, plastic or ferrite core.

Banana plug: A cylindrical connector approximately 4mm in diameter commonly used for multimeter test leads. Plugs into a banana socket.

Band: A range of radio frequencies allocated for a specific purpose or user, often by international convention and government allocation. As each band has different propagation characteristics to the ones either side major users such as broadcasters, defence and amateurs have multiple bands across the radio spectrum. See *sub-band*.

Band plan: An arrangement where amateurs choose their operating frequency to lessen interference to others, especially if using modes they cannot hear, read or decode. For example, you would avoid transmitting SSB in a digital modes segment or digital modes in a CW segment. Band plans are promulgated by national radio societies and are published on their website. They are mostly voluntary *gentlemen's agreements* except in the USA where the FCC regulates some band plans.

Band scope: A display on a screen, similar to a spectrum analyser, which allows you to watch the appearance of signals across an amateur band. Frequency is across the screen while signals show as columns with strength affecting height. A feature of software defined radio displays and in most middle and high-end HF transceivers.

Bandwidth (antennas): Refers to how much of a band you can operate with your antenna before the VSWR becomes unacceptably high. Antennas that are full size can cover most or all of a band while compact antennas (such as magnetic loops) can only have a narrow bandwidth unless efficiency is severely compromised.

Bandwidth (transmitting modes): Refers to the amount of spectrum a signal takes up. CW and digital modes have the narrowest bandwidth (a few tens of hertz), followed by SSB voice, AM voice, FM and then wideband data and television modes. Narrow bandwidth modes are more effective for a given power but have a slow data rate. Wide bandwidth modes the opposite; less efficient but a higher data rate. All but the cheapest receivers have bandwidth adjustments to suit the mode being received.

Baofeng: Chinese manufacturer of communications equipment. Most known for their low priced VHF/UHF handheld transceivers. http://baofengtech.com

Barefoot: Operating a transceiver without an external RF power amplifier.

Base: A connection on a bipolar junction transistor. In a common emitter amplifier configuration, the base corresponds to the input connection while the collector corresponds to the output connection. In this configuration a small base – emitter current flow (or signal) causes a larger collector – emitter flow (or signal).

Base loaded: Refers to a shortened vertical antenna with a loading coil near its feedpoint at the base of the antenna. Base loading is mechanically better but delivers inferior performance compared to centre or top loading.

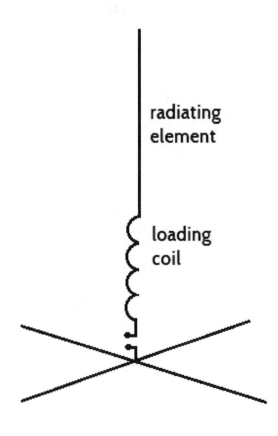

radiating
element

loading
coil

Base station: A non-amateur term used to refer to a station at a fixed location (e.g. home) as opposed to mobile or portable. May also refer to a large transceiver intended for home use. Unlike mobile or portable equipment, such transceivers often have inbuilt power supplies for connection to the AC mains.

Battery: A set of series connected cells used to power portable equipment. Main types are primary (non-rechargeable) and secondary (rechargeable) batteries.

Battery capacity: Refers to the amount of current or power a battery can deliver over time before it needs to be recharged. Expressed in amp hour (AH) or watt hour (WH). Higher capacity batteries operate equipment for longer but are larger, heavier and dearer. For example, a 5 AH battery can operate a receiver that draws 0.5A for 10 hours, whereas a 1 AH battery could only run it for about 2 hours.

Baud: A unit of data transmission speed. Equivalent to bits per second in digital communications.

Baycom: A popular 1990s computer program that allowed packet radio communication with a simple modem, computer and transceiver.

BBS: Bulletin board system. A packet radio station (comprising of a computer, software, modem and transceiver) that can store and forward messages. The storage function allows recipients to retrieve them later (similar to email) while forwarding allows them to be accessed from the BBS closest to the recipient. BBSs were often linked to form a worldwide packet radio network.

BCI: Broadcast interference. Interference caused to an amateur receiver from a nearby broadcast station. Improved front-end filtering in the receiver often resolves this.

BCNU: Be seeing you. (CW abbreviation)

Beacon: A continuously on transmitter that provides a known signal useful to test receivers or the presence of radio propagation. Typically erected by radio clubs or individuals, beacons are often located on tall buildings or hill tops.

Beam: A directional antenna, such as a yagi or quad, which concentrates the signal in one direction at the expense of others. This concentration provides a gain that boosts signal while nulls reduce noise and interference from other directions.

Beamwidth: The width of the main lobe of a beam antenna. A longer, higher gain beam typically has a narrower more concentrated beamwidth than a small beam with only two or three elements. A narrower beamwidth requires more careful rotation than a beam with a wide lobe.

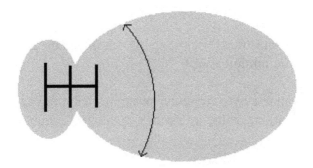

lower gain, wider beamwidth

higher gain, narrower beamwidth

Beehive trimmer: A type of air-spaced trimmer capacitor. With a maximum capacitance of 25 or (less commonly) 50 pF they are most used in VHF equipment.

Beta: The DC current gain of a transistor (β).

Beverage: A long low to the ground wire antenna often used for low and medium frequency reception due to its low noise pick up. The inclusion of a resistor to earth at its end makes the beverage directive, further reducing noise.

antenna wire typically 1 to 4 wavelengths long

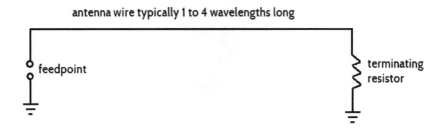

feedpoint

terminating resistor

BFO: Beat frequency oscillator. A circuit used in a radio receiver in conjunction with a mixer or detector. Can be added to AM only receivers to allow reception of CW and SSB signals. Or it might provide a needed signal (or 'beat note') for a receiver product detector to convert the incoming radio frequency signal to audio.

Bias: A small DC voltage or current applied to a part of a circuit so it can operate properly. For example, in a transistor linear amplifier bias is often applied to its base or gate connection to set its correct operating point and allow signals to be faithfully amplified. Too little or too much bias can cause insufficient amplification, excessive amplification, distortion, over-heating and even component destruction.

Bidirectional (antenna): Relates to an antenna with a radiation pattern favouring two directions, for instance a horizontal half wave dipole or single quad loop.

Bidirectional (circuits): Relates to stages in a receiver or transceiver that can be made to operate with signals arriving in both directions. This simplifies circuit design and signal routing. A recent example is the well-known Bitx series of HF transceivers.

Bifilar inductor: An inductor comprising two wires that have been bound or twisted before being wound on a toroid. Common in balun transformers where a 4:1 impedance transformation is required.

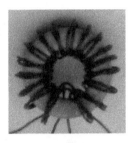

Big 3: The major three Japanese manufacturers of amateur transmitting equipment. Yaesu, Icom and Kenwood.

Bit loss: Corruption or degradation in a digital signal due to interference or a poor path.

Bitx: A popular homebrew single band SSB transceiver developed by Ashhar Farhan VU2ESE. http://www.hfsignals.net

BJT: Bipolar junction transistor. Standard transistor used in many electronic circuits. Has base, emitter and collector connections. Available in two main families – NPN and PNP. NPN has the collector more positive than the emitter. PNP has the emitter more positive than the collector. Other differences between BJTs include small signal transistors versus power transistors and audio versus RF (with the latter having a higher frequency cut-off limit).

BK: Break. A pause during a transmission. (CW abbreviation)

Black box operator: An amateur who uses all commercially built equipment because their greatest interests lie in operating. Opposite to *homebrewer*.

Black out: An interruption to ionospheric radio propagation caused by radiation from solar flares ionising the D-layer. This absorbs radio signals and degrades propagation.

Bleed over: Relates to a signal so broad that it is heard on frequencies that it shouldn't (e.g. 5 or 10 kHz up the band). This may be the fault of a wide or over-driven transmitter or a receiver that is either unselective or prone to strong signal overload.

Bleeder resistor: A resistor connected across a capacitor in a high voltage power supply to ensure that it is discharged when power is removed. Another application is across the input of a wire antenna to ensure that atmospheric static is shorted to earth.

Block diagram: A depiction of the circuit stages in equipment (such as a transceiver) in simplified form without identifying individual components. Each block is a stage or module that performs a particular function.

Blocking capacitor: A capacitor in a part of the circuit that requires DC current to be blocked but AC to pass. An example is between successive audio or RF amplifier stages where you wish to individually control the DC bias conditions of each transistor.

BNC: A popular bayonet-style connector for transceiver antenna leads. Valued for its small size, quick connection/disconnection and performance at UHF frequencies.

Board only: Refers to a short-form kit that contains the circuit board and possibly instructions. You have to find the remaining parts yourself. Some board-only kits may also have surface-mount parts pre-soldered.

Boatanchor: Term for an old and heavy piece of radio gear, normally employing vacuum tubes. Often lovingly restored by vintage radio enthusiasts and others who wish to recreate their childhood.

Bobtail: A type of vertically polarised wire curtain antenna used on the lower frequency HF bands. Popular amongst DXers operating on 1.8, 3.5 and 7 MHz.

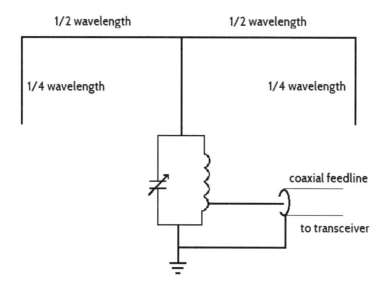

Body corporate: An association of home owners that administers common property in strata title residential development. May have regulations with regards to the erection of antennas that restrict amateur activities. (Australia)

Boom: The horizontal support for elements in a beam antenna. May be metal, fibreglass or wood.

BP: Bipolar. Often marked on capacitors. Denotes a special type of electrolytic capacitor that can be connected either way around.

BPF: Band pass filter. A filter designed to pass signals over a range of favoured frequencies. Signals on frequencies below or above the pass band are attenuated severely.

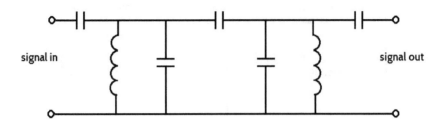

Braid: Woven wire mesh used as the outer or shield conductor on coaxial cable. Good cable has tightly woven braid while poor cable has thin loose braid that offers incomplete coverage of the inner conductor.

Breadboard: A form of construction used in the early days of radio based on screwing radio parts onto a piece of timber ("breadboard").

Break-break: A term often used by those who wish to urgently join a radio conversation. Properly used by those who have an urgent or important message but improperly used by the impatient.

Breakdown voltage: The minimum voltage above which a component or insulator 'breaks down' and starts to conduct. Sparks ensue and, unless current is limited, the part self-destructs.

Break in: To join a contact already in progress by announcing your callsign in gaps between transmissions. Announcing 'break' or (especially) 'break break' is more abrupt and not encouraged, especially if you were not previously involved in the contact or your message is not urgent.

Break-in: Refers to a transceiver feature or setting where you can listen between letters when transmitting Morse code. This is handy for high speed conversing, DXing, or message handling. Also known as QSK.

Breaking up: Said of a voice transmission that is hard to read such as might be caused by a faulty microphone connection, poor signal into a repeater or distorted audio. On SSB the term often implies a transmitter fault, as opposed to the signal merely being weak.

Bridge: A circuit configuration widely used for electrical measurements. Bridge circuits often includes fixed value components of known value, a connection to an unknown component, a known value component that is made variable and a sensitive voltage indicator such as a meter movement. A signal is applied to the bridge and the variable component adjusted until the meter reads 0. The bridge is then said to be balanced. A different value unknown component will require resetting the variable component to re-establish the null. A dial attached to the variable component then allows the unknown component's value to be read if the instrument has been calibrated with components of known value. The bridge circuit below measures resistance when supplied with a DC current but the same concept works for capacitance and inductance if an AC signal is applied and the meter circuit modified.

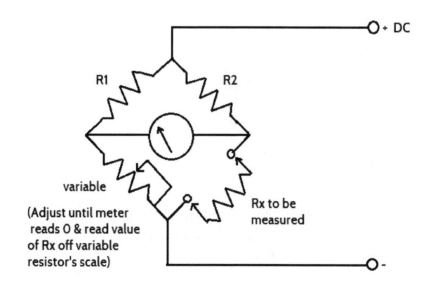

Bridge rectifier: A package of four diodes that allows AC to be converted to DC such as required in a power supply circuit. Bridge rectifiers can also be used to make DC powered equipment immune from reverse polarity connection. The diodes may either be separate or mounted in a package. Also known as a full wave bridge rectifier.

Broadband: Of or having a wide bandwidth. Examples include high-speed data transmissions or an antenna useful over a wide frequency range. Also wideband. Opposite to *narrowband*.

(the) Broadcast: A typically weekly news bulletin of amateur radio news produced and aired by a radio society or club. Normally followed by callbacks. (Australia)

Broadcast band: Refers to either the MF AM (or less frequently) the VHF FM frequency spectrum used for commercial radio broadcasting.

BTU: Back to you. Sent to warn your contact that it's their turn soon. (CW abbreviation)

Buffer: A circuit stage that isolates the VFO in a transmitter from subsequent stages. This is good radio design practice to minimise frequency drift and chirp.

Bug: An old type of semi-automatic side-to-side keyer for sending Morse code. Dits are automatically timed while dahs are manually timed.

Bureau: See *QSL Bureau*.

Buro: See *QSL Bureau*. (CW abbreviation)

Bypass capacitor: A capacitor used to provide a low resistance path for AC signals such as audio or RF while blocking DC currents. A common application is in a transistor audio or RF amplifier circuit in combination with a resistor where you need to regulate the DC operating point of the transistor by introducing resistance while allowing the AC signal to pass via the capacitor.

C

C: Refers to capacitors or capacitance.

C (operating): Yes. Confirmed. Correct. Confirms a callsign or message has been received correctly. (CW abbreviation)

Cabrillo: A common format for radio log data devised by Trey Garlough N5KO. Exported by logging programs when submitting a contest or award log online.

Calibrate: To adjust equipment, typically with the assistance of a frequency reference, so that its dial is reading the correct frequency.

Call area: Geographic division of a country used to allocate amateur callsign prefixes. In large countries call areas may comprise a province, state or group of states.

Call book: A directory of amateur callsigns, names and addresses. American callbooks are no longer published but they remain available for countries with smaller amateur populations.

Call sign: The unique collection of letters and numbers that identifies your amateur station on the air. Typically comprise a prefix (signifying your country) and suffix. Regulations specify that you send your call sign at regular intervals during a contact.

Calling frequency: A frequency that stations call to make contact and then move away from once contact has been established. Most common above 29 MHz, especially for SSB weak signal and channelised FM modes. Typically specified in *band plans*.

Capacitance: Refers to the measure of a capacitor's ability to store electrical charge. The basic unit is the Farad, but since this is a large unit, capacitance is more commonly expressed as microfarads, nanofarads or picofarads.

Capacitance hat: Refers to a horizontal 'hat' of stiff wire or metal tubing placed at the top of a vertical antenna to reduce its resonant frequency or make it more efficient.

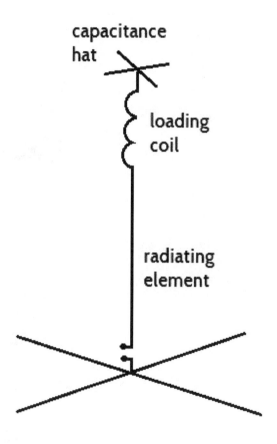

capacitance
hat

loading
coil

radiating
element

Capacitive reactance: Resistance to AC signals (including those at radio frequencies) caused by capacitance in a circuit. Unlike resistance, whose value is independent of frequency, the reactance of a capacitor decreases with frequency. And unlike inductance, a capacitor's reactance decreases with increasing value. Another property of capacitance reactance is that its effect can be cancelled out by inductance, but only for signals of a particular frequency. This frequency dependent behaviour is what makes tuned circuits work.

Capacitor: An electronic component that stores electrical charges and can pass AC (but not DC) currents and signals. Capacitors are useful for a wide range of coupling, filtering and oscillator circuit applications.

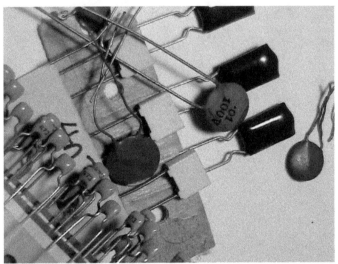

Call back: A session sometimes held after a broadcast or net where listeners confirm that they have been listening and can offer any comments.

Capture effect: A characteristic of FM radio communication that causes only the stronger signal to be heard if two stations transmit simultaneously on the same frequency.

Carbon film resistor: All-purpose cylindrical resistor available in power ratings up to 2 to 5 watts. Values vary from under 1 ohm to 10 megohm, indicated by striped colour code on body.

Carbon microphone: An old-fashioned type of microphone where sound causes carbon granules to vibrate, varying their electrical conductivity. Used in many telephone handsets and radio transceivers up to the 1970s.

Cardioid: Refers to a heart-shaped, broadly unidirectional pattern. Examples include the radiation pattern of certain configurations of phased vertical antennas or directional microphones. Such a pattern reduces the pick-up of noise from unwanted directions.

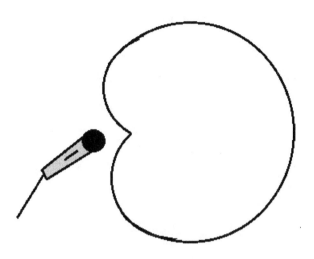

Carrier: An unmodulated radio signal such as produced when pressing a microphone's push to talk button on an AM or FM transmitter. A carrier can be interrupted to send Morse code (CW).

Carrier balance: An important setting in an SSB or DSB transmitter that sets the balanced modulator so that it nulls out any carrier signal that may be present. Misalignment can mean that your signal contains an undesirable carrier.

Carrier dropper: A person who transmits a carrier without identifying, either carelessly when tuning up or deliberately to cause interference.

Carrier oscillator: A stage in an SSB transmitter that generates a carrier signal, only to be later suppressed in the balanced modulator. The resultant double sideband suppressed carrier signal may be fed to a crystal filter to be made single sideband before being converted to the desired operating frequency in a mixer stage.

Carrier suppression: The extent to which the unwanted carrier signal is suppressed in a DSB or SSB transmitter. Measured in decibels relative to the transmitter's maximum RF power output, a suppression of 50 or 60 dB down is considered good.

Cathode: Connection on a diode or electron tube that current flows out of (according to conventional current flow – electron flow is the opposite direction). Opposite to anode.

Cat's whisker: A type of detector used in early crystal radio receivers. Employed a wire that was touched on to a sensitive spot on a lump of galena to detect signals. Since replaced by germanium diodes.

Cavity filter: A high quality radio frequency filter, such as used in an amateur radio repeater, to prevent the receiver being desensitised by the strong, same-band signal generated by the repeater's transmitter.

CB: Citizens Band. Short range low power radio communication open to everyone without having to pass a test or apply for a license. Most countries have CB allocations around 27 MHz. CB's limitations have encouraged many wishing to take their radio interest further to become licensed amateurs.

CC&R: Covenants, Conditions and Restrictions. Petty regulations imposed on property owners by home owners associations (HOAs) in some housing developments. These may restrict the colour residents can paint their homes, the style of any fences and (important for us) whether outside TV and radio antenna masts can be erected. Amateurs desiring the freedom to put up good antennas are encouraged to buy homes not subject to CC&Rs. Or, if living in a CC&R development is unavoidable, become friends with neighbours and get on the home owners association committee to reach agreement on erecting antennas. (USA)

Cell: A section of a battery. Depending on type, cells typically have a voltage between 1.2v and 3.7v each, so several need to be connected in series to provide a useful voltage for radio equipment.

Centre fed: Refers to a wire antenna, such as a common half wave dipole, that has the feedline connected to its centre. The feedline can either be coaxial cable (either directly or via a balun) if one band is required or open wire feedline if multiband coverage is desired.

Centre loaded: Refers to a vertical antenna with a loading coil about half way up the antenna. Centre loading is mechanically inferior but delivers better performance than base loading.

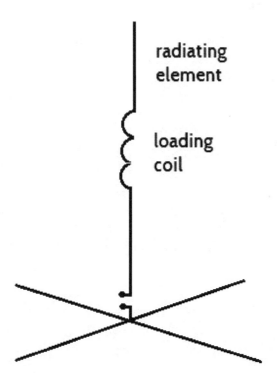

CEPT: Conference of Postal and Telecommunications Administrations. In amateur radio usage refers to licensing arrangements that allow visiting amateurs to operate outside their own country without paperwork. European based with some participation from countries outside Europe. http://www.cept.org

Ceramic filter: A type of band pass filters often used in transmitter or receiver circuits to provide selectivity.

Ceramic resonator: A component, similar to a crystal, which can be made to oscillate at a specific radio frequency.

Ch: Channel. A frequency allocated for radio communication. May also be a frequency that is part of a channelised band plan.

Channel spacing (or channel step): The interval between adjacent channels in a channelised frequency allocation system such as used for VHF and UHF FM and repeater communication. Common channel spacings include 12.5, 15, 20, 25 or 30 kHz. Channel spacing is typically set to exceed a signal's bandwidth so that adjacent channels can be used without causing bleed over to adjacent channels. Most equipment offers a choice of channel steps to coincide with channel spacing used for various communications services.

Channelised: Refers to segments of bands where operators through law or custom transmit on specific frequencies only. Examples of channelisation enforced by law is the 27 MHz CB radio band and UHF FRS or PMR bands. The FM and repeater segments of VHF and UHF amateur bands are examples of channelisation by custom.

Characteristic impedance: The ratio between voltage and current of a radio signal being carried along a uniform transmission line in the absence of reflections in the other direction. Common values include 50 and 75 ohm for coaxial cable and 300 or 450 ohm for ladder line.

Chaser: A person who seeks to contact a sought-after station (see activator) on the air, notably for an awards program such as DXCC, IOTA or SOTA.

Chassis: The main internal subassembly used to hold circuit boards or components inside a piece of electronic equipment. Metal chassis are often found in older equipment. The chassis is often (but not always) connected to earth or ground.

Check in: To call in to a *net*.

Chewing the rag: Having a *ragchew*.

Chirp (Morse code signal): Refers to a fast frequency drift on characters when sending Morse code. Caused by insufficient buffering or poor power voltage regulation.

Chirp (software): Popular open-source software for programming settings and memory channels into a transceiver via a computer. It supports a wide range of transceiver models. http://chirp.danplanet.com

Chordal hop: An HF propagation mode based on signals being carried within the ionosphere for some distance before being reflected back to earth. Can allow long-distance communication with low transmit powers.

Chromapix: A common SSTV transmitting and receiving computer program. http://www.barberdsp.com

Circuit breaker: A fuse-like component that removes power from a circuit if excessive power is drawn, saving it from further damage. Unlike a fuse a circuit breaker can be reset once the problem is fixed.

Circuit diagram: A depiction of the parts used in a piece of electronic equipment with components drawn as electronic symbols. Also *schematic diagram*.

Circular polarisation: Neither horizontal nor vertical polarisation, a signal with circular polarisation is one whose polarisation rotates. There is both left and right-hand circular polarisation, depending on the direction of the electric field's vector rotation. Circularly polarised antennas are often preferred for satellite communication due to less fading on received signals compared to if conventional vertical or horizontally polarised antennas were used.

CL: Clear. Closing my station. See *Clear*. (CW abbreviation)

Clarifier: Another term for fine tuning or RIT. In a transceiver the clarifier allows fine adjustment of the receive frequency to provide clearer reception of a signal without changing the transmit frequency. The ability to make small adjustments is particularly important for SSB reception where even a small frequency change greatly affects readability.

Class A: Refers to an inefficient but highly linear type of amplifier used for high quality audio and radio frequency amplification. The active device conducts over the entire range of the input cycle.

Class AB: A compromise between Class A and Class B in amplifier circuits, offering intermediate levels of efficiency and output signal quality.

Class B: A moderately efficient form of amplifier configuration suitable for linear audio or radio frequency amplification. There are two active devices with each conducting over one half of the input cycle.

Class C: Refers to a highly efficient type of RF power amplifier that is non-linear. In other words, the output power is not proportional to the input power applied. Such amplifiers are suitable for CW or FM transmitters. However, they are unsuitable for audio and modes such as SSB which requires a linear (e.g. Class A or B) amplifier.

Class D: An extremely efficient type of amplifier, employing pulse width modulation, whose amplifying devices operate as switches. The high efficiency means that smaller heat sinks are used than what would be required for other types of amplifier.

Class E: Another extremely efficient form of amplification used in some modern radio transmitters. The high efficiency is obtained with careful selection of component values to take advantage of amplifying device capacitances.

Class H: An efficient type of amplifier based on modulating the voltage of its power supply rail in sympathy with the input signal.

Clear: Going clear. Off and clear. Closing my station. Announced during the final transmission of a voice contact to signify that you will be switching off and not responding to subsequent calls.

Closed circuit: Refers to a connection, circuit or pair of switch contacts through which electrical current can flow.

Closed repeater: A repeater whose sponsors discourage usage by people outside their club or clique. Supporters say that those who fund maintenance are entitled to restrict access. Opponents contend that restricting access has no legal status and is against amateur radio's open ethos. (USA)

Club: An amateur radio club. May be geographically or special interest based. Typically holds meetings, runs exams, maintains repeaters, participates in contests and has social outings. Often affiliated with a national society.

Cluster: See *DX cluster*.

CMOS: Complementary Metal Oxide Silicon. A type of low-current integrated circuit that can be used over a range of voltages. Widely used in logic circuits. Has largely replaced the older current-hungry TTL chips.

Coax: Coaxial cable. A cable commonly used as feedline between a transceiver and antenna. Comprises an inner wire, plastic dielectric, braid and outer jacket. An unbalanced feedline. Thick coaxial cable is dearer but has less loss than thinner coax so is recommended for VHF and UHF antenna installations. Examples of coaxial cable include RG58, RG8 and RG213.

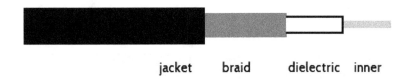

jacket   braid   dielectric   inner

Cobweb: A type of multiband HF antenna typically used on the higher frequency HF bands. Comprises parallel dipoles bent into a horizontal square and supported by fibreglass poles. This is attractive for homes with small square yards. Radiation pattern is approximately omnidirectional.

Coil: An *inductor*.

Collector: A connection on a bipolar transistor. It is the most positive connection with respect to ground for an NPN transistor and the most negative connection with respect to ground for a PNP transistor. In a common emitter amplifier configuration, the collector is the connection from which the amplified output is taken. It may then be fed to another stage, the antenna (if a transmitter) or headphones (if a receiver).

Collins: Respected US manufacturer of radio communications equipment. Most well known for their pioneering SSB transceivers.

Colour code: A system using coloured stripes on a component's body to indicate its value. Most common with resistors but also used with some inductors.

Common: Refers to a connection or stage of a circuit that is shared. For example, an earth connection may be common to several components or items of equipment. Or in a transceiver the VFO stage may be common to both the transmitter and receiver portions and control both.

Communications quality: Refers to voice transmissions whose audio characteristics have been shaped to maximise intelligibility through fading and interference. Typical characteristics of a communications quality signal include compressed audio (to bring up quieter syllables) and audio tailoring (to boost intelligibility-giving voice frequencies around 2-3 kHz).

Complex impedance: An impedance that is not purely resistive such as encountered with antennas used away from their design frequency. A complex impedance contains both a resistive and reactive component (which can be either capacitive or inductive). An antenna with a complex impedance cannot be efficiently used with a transceiver unless its complex impedance can be transformed to 50 ohm resistive with some form of matching network or antenna coupler. And even then you need to be aware of potentially increased feedline loss if the matching network is at its transceiver end.

Compression: Refers to signal processing that reduces the variation between its strongest and weakest parts. Achieved through a special amplifier stage, called a compressor, which uses feedback to reduce gain on strong signals and increase gain on weak signals. Limited speech compression on a transmitter can improve its ability to punch through when signals are weak. However excessive compression reduces readability and boosts unwanted background noise.

Compression trimmer: A type of adjustable trimmer capacitor comprising two plates separated by mica insulation. Turning the adjustment screw compresses the plates and increases capacitance. Can provide a large capacitance from a small space such as required in MF and HF radio equipment.

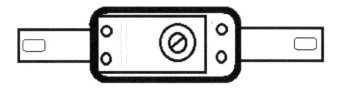

Conductor: Material that has a very low resistance to the flow of electrical current (e.g. most metals). Examples of conductive metals used in electronic applications include copper, tin, brass, silver and gold. Aluminium is also sufficiently conductive for equipment chassis and antenna elements but its tendency to oxidise and difficulties with soldering limit its use.

CONDX: Conditions. Propagation conditions. Normally described as either poor or good. (CW abbreviation)

Constant voltage: Refers to a battery charger or power supply that has a regulator to keep the voltage constant regardless of changes in current draw. This is required to power stable oscillators and amplifiers or charge certain types of battery such as lead-acid.

Constant current: Refers to a battery charger that has circuitry to maintain a steady flow of current into the battery it is charging. This is suitable for charging nickel-cadmium and nickel metal hydride batteries.

Contact: A conversation made via amateur radio. Also *QSO*.

Contest: A scheduled on-air competitive event where amateurs earn points by making as many contacts as possible within a given time period. Major contests are held annually and run for 24 or even 48 hours. Smaller contests may be briefer (see *sprint*). Contests bring out many rarely active amateurs and fill the bands with activity.

Contest callsign: A special callsign available in some countries for use by radio clubs or groups for amateur radio contesting. This can give an advantage where the callsign is shorter or quicker to send.

Continuously tuneable: Refers to radio equipment that can operate on any frequency in a band, i.e. not restricted to specific channels. Examples of tuneable equipment are HF amateur transceivers using either a free-running VFO or frequency synthesiser with fine tuning steps. Opposite to *channelised*.

Convention: An organised gathering of hams, often featuring talks, displays and equipment sales. Similar to a hamfest.

Conventional current flow: The concept that electrical current flows from positive (+) to minus (-) in a circuit as used by most radio and electronic theory books. This is opposite to electron flow which is from negative to positive.

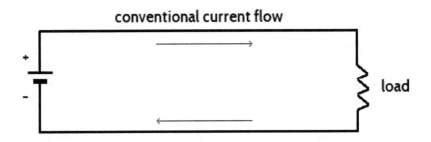

Convert: The act of modifying a transceiver intended to operate on one band (often non-amateur) so it can operate on another (often amateur). In the 1970s and 1980s 27 MHz CB radios were often converted to the 28 MHz amateur band. Similarly, VHF or UHF FM transceivers could be converted to the 2m or 70cm bands.

Converter: A circuit comprising an RF oscillator, RF mixer and band pass filters that shifts frequency up or down through addition or subtraction. Found inside superhet receivers or can be added externally to extend the tuning range of a receiver. A transverter does similar for a transmitter or transceiver.

Cooling fan: A fan installed on a computer, transmitter or power amplifier to dissipate heat generated by internal circuitry. A faulty cooling fan can cause overheating and more serious problems to occur.

Complex impedance: Refers to an impedance that is not purely resistive and thus needs to be expressed as a two-part (i.e. complex) number. Complex impedances have two components – a resistive component and a reactive component, which can be either capacitive or inductive. The reactive component (j) is expressed as an imaginary number with – indicating capacitance and + indicating inductive reactance.

Copy: To understand a radio transmission.

COR: Carrier operated relay. A relay that activates when a radio signal has been received, such as required to cause a repeater to transmit. The circuitry that drives the relay is typically connected to the receiver's squelch line that becomes active when a signal is present.

Corner reflector: A V-shaped arrangement of reflector elements located behind a driven element to form a directional antenna used at VHF and UHF.

Cornflake (or cereal) packet: A pejorative term used to refer to the alleged source of someone's amateur licence, especially if they are considered to have the technical knowledge of a *screwdriver expert* or the operating skills of a *lid*.

Corona discharge: A light-giving electrical discharge caused by the ionisation of air around an electrically charged conductor. Coronas may sometimes be seen around high voltage points of antennas connected to high power radio transmitters. It is desirable to eliminate corona discharges from antennas, for example by making the ends of antenna elements less sharp.

Counterpoise: Wire or wires often used as an earth or ground substitute in an antenna system. Certain types of antennas need to be loaded against a counterpoise to be efficient and minimise unwanted radiation from the feedline.

Coupling: The transfer of energy or signals from one part of a circuit to another. May be conductive, capacitive or inductive. Coupling can be desired or undesired, with the latter ('stray coupling') often interfering with a circuit's performance.

Coupling capacitor: A capacitor used to provide a low resistance path for AC signals such as audio or RF while keeping circuits isolated from one another as far as DC is concerned. A common application is between two audio or RF amplifier stages.

Co-sited: Refers to two or more repeaters, beacons or other facilities that share the one site.

Coverage: Refers to the normal distance that a radio transmitter or station can transmit over. Affected by factors such as transmitting frequency, transmitter output power, antenna gain, height, terrain and location.

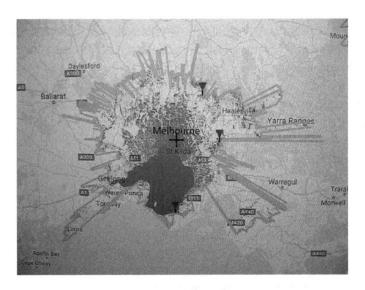

CPO: Code practice oscillator. An audio oscillator connected to a key used for practicing Morse code.

CQ: A general call to any station. Used when you wish to establish contact with anyone anywhere. One of the finest traditions of amateur radio.

Calling CQ epitomises the open amateur ethos and is what makes our interest distinctive from all other forms of electronic communication; unless we're telemarketers or spammers we don't phone random numbers or send random emails for example.

Instead of speaking to the same circle of people on a net, calling CQ shakes things up a bit so you're hearing and working different people. Even better is if you find you have something in common and can get past the 'rubber stamp' name, location and signal report.

Wouldn't it be better if there were no directional calls, skeds or nets on one day of the week and all who would have come on called CQ or tuned the band instead? One might be surprised at the contacts made.

There are good and bad ways of calling CQ. Firstly, study the band plan for your particular mode and listen around. If no other stations are calling CQ find a likely frequency, ask if it's in use and commence calling. Calling should be slow and steady with your callsign pronounced clearly and repeated. Don't give up after two or three calls – it might take 15 or even 30 minutes to get a contact on the quieter bands.

CQ magazine: A US-based amateur radio magazine. Less technical than some others, it emphasises the operating, DX and contesting facets of the hobby. http://cq-amateur-radio.com

CQ WPX: CQ WPX Contest. A popular international transmitting contest sponsored by CQ magazine. Based on working as many callsign prefixes as possible. Its CW and SSB sections are held in the first half of the year.

CQ WW: CQ Worldwide Contest. A popular international transmitting contest sponsored by CQ magazine. Its separate SSB and CW events are both held late in the year.

CQ zone: Refers to a system of dividing the world into 40 geographical zones for the purpose of particular awards and contests sponsored by CQ magazine.

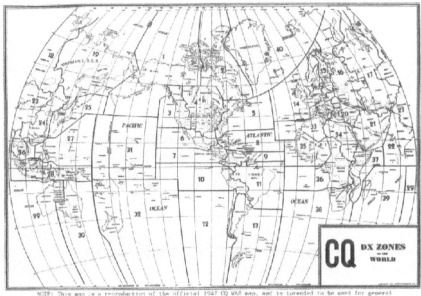

NOTE: This map is a reproduction of the official 1947 CQ WAZ map, and is intended to be used for general reference only. Over the years numerous adjustments have been made to Zone boundaries for various reasons. Always consult the latest CQ WAZ Rules for the most up-to-date Zone delineations.

Crimp connector: A solderless style of electrical connector, terminal or plug. Requires a special crimping tool to provide a firm connection. The merits of solder versus crimp connectors are often hotly debated amongst hams.

Crimping tool: A tool used with crimp connectors to form crimp connections – that is solderless electrical connections based on pressure between a wire and a connector.

Critical frequency: The highest HF frequency that the ionosphere will reflect signals straight down such as required for NVIS (or short skip) communication. Tends to be lower during low sunspot years and at night. Frequencies below the critical frequency are suitable for blanket communication up to several hundred kilometres. Whereas transmissions between the critical frequency and the MUF will skip over closer in stations but span longer distances well.

CRO: Cathode Ray Oscilloscope. Also oscilloscope. A piece of test equipment that allows the monitoring of audio and radio frequency signals, such as may emanate from an audio amplifier or radio transmitter. Previously contained a cathode ray tube for the screen, hence the name.

Cross-band: Amateur communication involving transmitting and listening on different bands. The full-duplex communication possible allows free-flowing conversations similar to a telephone as listening stations can interrupt talking stations.

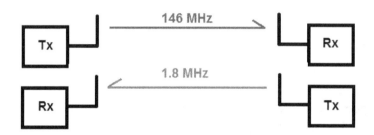

Cross-polarisation: Refers to communication where one station is using a horizontally polarised antenna and the other is using a vertically polarised antenna. Under some circumstances (e.g. local VHF and UHF contacts) using different antenna polarisations greatly weakens signals and shortens communication range. Also see *polarisation*.

Crowbar: A protection circuit in a power supply that shuts it down should a malfunction cause an excessive output voltage. This is to protect the equipment connected to it.

CRT: Cathode Ray Tube. A type of vacuum tube for the display of images such as used on television sets, computers and test equipment up to the 2000s.

Crystal: Quartz crystal. A component that when placed in a transistor or IC oscillator circuit allows a stable signal of accurate frequency to be generated. Crystals can be used on their own in simple fixed frequency transmitters or used to generate a stable reference signal for a frequency synthesiser. Another application is to pass a narrow range of frequencies, such as with a *crystal filter*.

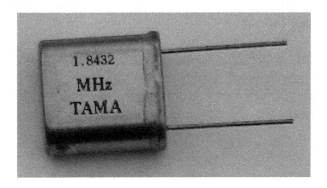

Crystal calibrator: An item of test equipment containing a crystal oscillator on a known frequency followed by frequency dividers. This can be used to establish a receiver's exact frequency, especially if it's the older type without a frequency synthesiser and digital frequency display.

Crystal controlled: Refers to a transmitter or receiver that can only operate on one or more fixed frequencies regulated by a crystal oscillator. Most common for pre-1975 VHF equipment and simple HF AM and CW transmitters. Opposite to *frequency agile*.

Crystal diode: Germanium diode. Type of diode that starts conducting at a low voltage e.g. 0.3v. This makes it useful for certain applications such as crystal set receivers and test equipment.

Crystal earpiece: A high sensitivity high impedance earphone using the piezoelectric effect. Commonly used with crystal set receivers.

Crystal filter: A number of quartz crystals connected so that they pass only a narrow bandwidth such as that occupied by a CW, FM or SSB signal. Widely used in transmitters and receivers to allow SSB to be generated and provide receiver selectivity.

Crystal grinding: An old technique where quartz crystals were taken apart and ground down with grit to raise their frequency slightly. This was common if you needed an amateur frequency from a crystal just outside the band.

Crystal microphone: A high impedance microphone that generates a small signal through the piezoelectric effect. Known for its sharp 'communications quality' audio they were popular in the 1960s and 70s.

Crystal oscillator: An IC or transistor RF oscillator using a quartz crystal to regulate its output frequency. Can be built from discrete components or available in some frequencies as a module.

Crystal radio: Crystal set. A radio receiver entirely powered by the incoming signal. Requires no external power. Popular in the early days of radio and as a beginner's project.

CTCSS: Continuous Tone Coded Squelch System. A system used on VHF/UHF FM communications systems that requires a low-pitched tone to be impressed on a transmission for it to unmute the receiver. This is handy for commercial applications where one frequency is shared by several groups of users and it would be annoying to hear other peoples' conversations. Amateurs also use CTCSS for purposes such as activating crossband repeater links or lessening false triggering from stray RF sources.

CU AGN: See you again. (CW abbreviation)

Cubical quad: A two element quad beam antenna with two square wire loops supported by spreaders. Comprises a driven element and either a reflector or director. So-called as its outline resembles a cube.

CUL: See you later. (CW abbreviation)

Current balun: A type of balun that forces equal (but opposite in phase) currents to each of its balanced side connections (assuming a signal is being applied to its unbalanced side). Commonly used in antenna work.

Current limiting: Refers to a feature in a power supply that limits the current delivered, even if a short circuit is connected across its terminals. This is particularly useful when servicing as it can prevent further harm to already faulty equipment.

Curtain array: A class of fixed wire high gain directive antenna often used by HF broadcast and other stations that need to concentrate their signals in a particular direction. Despite their good long-distance performance, they are rarely used by amateurs due to the height of masts and amount of land required.

CW: Continuous wave. Conventional Morse code radio communication involving a carrier signal interrupted by a key. A receiver with a beat frequency oscillator or regenerating detector is required to hear the signal.

C4FM: Constant envelope four level frequency modulation. A type of frequency shift keying used for digital voice and data modes such as System Fusion.

D

D-region: The first significant layer (or region) of the ionosphere closest to earth. Its maximum concentration is around 60 to 90km altitude. D is for destructive and in this case the layer's main effect is to absorb radio signals in the low and medium frequency parts of the spectrum. The D layer is present during the day and disappears around dusk. Daytime D-layer absorption is the main reason why the medium frequency AM broadcast band is alive with distant stations at night but not during the day.

Data mode: A mode based on sending text data, normally in digital form and involving a computer. Examples include packet radio, PSK31, JT65, Olivia, FT8, JS8 Call, WSPR and more, with new ones being added and refined all the time. Different modes suit different purposes, such as manual keyboard chatting, 'rubber stamp' contacts and weak signal beaconing. All can be accomplished with a transceiver, interface box, computer and special software. Alternatively, you can buy stand-alone transmitters for modes such as WSPR that free your transceiver and computer for other tasks.

Data modules: Refers to small UHF transmitter and receiver modules that can be used to transmit data (or even, if modified, voice) over short distances. Due to their low power a licence is not normally required to use them.

DATV: Digital Amateur Television.

dB: Decibel. One tenth of a Bel (which is rarely used). A ratio commonly used to compare signal strengths, antenna performance and amplification ratios in radio and electronics. It is a logarithmic scale. A change of 3dB (equivalent to a doubling of power) is barely noticeable while a 10dB change (equivalent to x10 power) is very noticeable. A table comparing dB to power ratios is below:

-10dB	x0.1
-7dB	x0.2
-5dB	x0.33
-3dB	x0.5
-1dB	x0.8
0dB	no change
+1dB	x1.25
+3dB	x2
+5dB	x3
+7dB	x5
+10dB	x10
+13dB	x20
+16dB	x40
+20dB	x100
+30dB	x1000

dBd: Antenna gain in decibels relative to a half wavelength dipole. Term used to compare the gain performance of other antennas such as beams. Since a dipole does not radiate evenly in all directions it has a small gain over an isotropic radiator (2.15dBi).

dBi: Antenna gain in decibels relative to an isotropic radiator (that is a theoretical antenna that radiates equally in all directions). E.g. an antenna measuring 5dBi gives a signal 5 dB stronger than would be the case if the transmitter was connected to an isotropic radiator. dBi is often used by antenna manufacturers as it makes gain figures look higher compared to if they were expressed relative to a half wavelength dipole (dBd).

DC: Direct current. Unidirectional flow of electric charge such as supplied by batteries.

DC to DC converter: A power device that converts DC current from one voltage to another. They may step up or step down voltage. Modern types use a switching regulator for small size and high efficiency.

DCS: Digitally coded squelch.

DDS: Direct digital synthesiser (or synthesis). A method of generating a stable radio frequency signal such as needed in a radio transmitter or receiver. Now standard in modern equipment. Low-cost DDS modules are also available for homebrew projects.

DE: From. Used when calling a station. For example: <station called> de <station calling>. It may also be used at the start and finish of transmissions. (CW abbreviation)

Dead bug: Anarchic and fast form of electronic construction. Involves soldering components to each other and to the copper side of a piece of unetched printed circuit board material for support and grounding. 'Dead bug' saves drilling time and allows easy changes to circuits such as required when prototyping. So-called because ICs mounted this way resemble dead bugs.

Decay: Refers to the speed that an AGC or audio compressor circuit reverts to full gain after the finish of a strong input signal. A fast attack slow decay setting is often used to minimise sudden spikes in output level.

Decoupling capacitor: A capacitor that short circuits AC signals to ground such as required to suppress AC ripple on DC power supply rails.

Delta loop: A triangular wire loop antenna often used on HF. Normally 1 wavelength perimeter on its design frequency, delta loops can be built apex up (with one tall mast) or apex down (with two tall masts). Polarisation is either horizontal or vertical, depending on the feed point location along the wire. Adding a slightly smaller or larger loop element to the front or rear forms a directional beam antenna similar to a two-element quad (except for the shape of the elements).

Demodulation: The process where an incoming radio signal is converted from RF to audio as in the detector stage of a receiver. An audio amplifier boosts the demodulated output sufficient to drive a speaker or headphones.

Desensing: Refers to the overload suffered by a receiver from a nearby transmitter on a nearby frequency. This spoils reception of weak signals. Most notable with large contest stations or VHF/UHF repeaters. Cures include heavily shielded equipment, widely separated antennas, high quality receiver front-ends and narrow filters.

Deviation: The maximum swing in frequency of an FM transmission away from its carrier frequency when audio is applied. 88-108 MHz FM broadcast stations use wide deviation, known as wide band FM – WBFM, while VHF/UHF communication services (including amateurs) use narrow deviation (known as narrow band FM - NBFM).

DF: Direction finding. The art of ascertaining the direction a signal is coming from with a receiver and directional antenna. Done for navigation, interference location or sport (see *ARDF*).

Dial light: A small light globe or LED used to illuminate the frequency display or dial of electronic equipment.

Dielectric: The insulating material between a capacitor's metal plates. Usually some type of plastic but can also be air for physically larger types of variable capacitor.

Dielectric constant: The extent to which dielectric material in a capacitor, etc can store electrical energy (and thus increase capacitance) relative to free air (which has a dielectric constant of 1 at room temperature). Expressed as a ratio compared to air.

Digipan: Digital Panoramic Tuning. An early 2000s computer program used for transmitting and receiving digital modes such as PSK31 and PSK-63.

Digipeat: To extend packet radio transmitting range using another station as a digipeater.

Digipeater: Digital repeater. Apparatus that retransmits a packet radio signal so that messages can be passed beyond the range of the originating station. Reception and transmission occur on the same frequency.

Digital: Refers to circuitry, devices and communication where information is conveyed in 0s and 1s rather than a continuously variable analogue signal. The use of a small number of known values has advantages in communications, computing and signal processing. Opposite to *analogue*.

Digital voice: An umbrella term for various techniques for sending speech transmissions via digital, as opposed to analogue, means. Current examples include P25, DSTAR, Fusion, DMR on VHF/UHF and FreeDV on HF.

DIL: Dual in line. Refers to a component package with two rows of parallel pins such as common with integrated circuits. Also *DIP*.

Diode: Electronic component that conducts electricity in one direction only. Used for many applications including converting AC to DC, demodulating radio signals, generating voice transmissions, mixing frequencies, voltage regulation, reverse polarity protection and more.

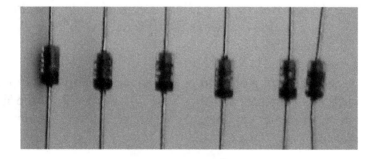

DIP: Dual in-line package. Refers to a package with parallel pins such as often used to house ICs. Also *DIL*.

Diplexer: A device that allows one antenna to be shared between two radio transmitters or receivers. An example could be a multiband 2m/70cm antenna used with separate 2m and 70cm transceivers.

Dipole: Most often a half wavelength of wire fed at the centre, this is a simple and efficient antenna suitable for operation on one band if fed with coaxial cable. Feeding with open wire feedline allows operation on multiple bands. Dipoles can be erected horizontally, vertically or as an inverted-vee.

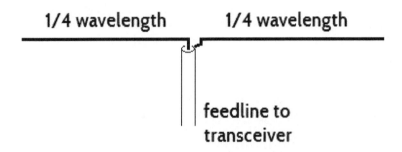

1/4 wavelength   1/4 wavelength

feedline to transceiver

Dip oscillator: An old-fashioned piece of radio test equipment that can measure the resonant frequencies of tuned circuits and act as a crude RF signal generator for receiver testing. Also known as a dip meter, grid dip oscillator or gate dip oscillator.

Direct (QSLing): Refers to mailing a QSL card directly to your contact's postal address. Faster than sending your card 'via the bureau' but more expensive.

Direct (repeater use): Refers to you hearing a station on a repeater input. If this is possible, and they can also hear you on the input then you can communicate directly without going via the repeater. See *simplex*.

Direct conversion: A type of receiver circuit where incoming signals are converted directly to audio in a mixer or product detector stage by being beat against a locally generated signal close to the frequency of that being received. Often simpler than a superhet which converts signals to an intermediate frequency before being converted to audio. Associated with simple beginner receivers, direct conversion techniques are now also used in software defined radios.

Direct wave: Refers to propagation of radio signals in line of sight as typical with VHF and UHF signals. Contrasts with ground wave (mainly LF and MF) and sky wave (mainly HF).

Directional: Relates to an antenna with a radiation pattern that favours one direction, such as a beam. Different to *bidirectional* (radiation favouring two main directions) and *omnidirectional* where the antenna squirts signals in all directions.

Directional call: Opposite to a general call (see CQ). Here you are calling either a specific station or specific group of stations, such as those in a particular country or participating in a shared activity such as a contest. A directional call to a specific station consists of their call sign followed by your own while a directional call for a particular event comprises a CQ followed by the event or contest name.

Directional coupler: An unpowered piece of equipment that allows a small amount of RF energy to be tapped off for RF measurement or monitoring purposes.

Director: Element in front of a driven element in a beam antenna. May be straight (for yagi antennas) or a wire loop (for delta or quad antennas). Normally shorter than the driven element. More directors typically mean a more directive beam with narrower beamwidth and higher gain. Compare with *reflector*.

Disc ceramic capacitor: A non-polarised capacitor used in applications where low capacitance is required such as in radio frequency equipment. They look like dry lentils. Typical values are 1 pF to 220 nF. NPO types are more stable in value over varying temperatures.

Discharge: Refers to the draining of stored electrical energy from a battery or capacitor.

Discone: A vertically polarised unity gain antenna able to cover a wide range of VHF and UHF frequencies. Comprises a horizontal disc above a cone pointing upward. To reduce weight and wind loading, both are normally formed from tubular element stock rather than being continuous material.

Discrete components: Separate parts such as resistors, capacitors, transistors, etc that perform a single function only. Contrasts with integrated circuits which have multiple components on the one chip.

Discriminator: A stage in an FM receiver that converts incoming frequency modulated signals to amplitude modulated signals, enabling demodulation.

Dish: A directional high-gain parabolic antenna commonly used on microwave frequencies.

Distortion: Undesirable change to the waveform of an audio or radio signal as it passes through a microphone, amplifier, speaker etc.

DMM: Digital multimeter. An instrument that can measure voltage, current and resistance. May have other features such as capacitance, inductance and diode testing.

DN: Down. Often with reference to moving to another frequency. (CW abbreviation)

Dogpile: A large number of stations transmitting at once, such as may occur when replying to a call from a sought-after *DX station*.

Doppler shift: Variation in frequency due to the movement of an object transmitting a signal through space towards or away from the receiver. This effect is most noticed by amateurs when communicating via orbiting satellites.

Double: To transmit while someone else is transmitting, typically as a result of a misunderstanding or signal fading. If done on SSB those listening hear both transmissions but may understand neither if signal strengths are similar. If done on FM the capture effect means that only the strongest signal is heard.

Double balanced mixer: A type of mixer that provides high isolation between all input and output connections as required for high performance receivers and transmitters. This isolation minimises leakage of signals on either of the mixer's inputs through to the mixer's output.

Double conversion: See dual conversion.

Double extended zepp: A long centre-fed doublet type antenna that is 1.25 wavelength from end to end. In other words, each side is 5/8 wavelength long. Such an antenna exhibits narrower lobes and higher gain than the more common half wavelength dipole (only 1/4 wavelength per side). It is normally fed via open wire feedline and a balanced antenna coupler.

Double insulated: Refers to an electrical appliance with two layers of insulation between its live parts and the outside world. Such insulation allows it to be safely used without having an earth connection. Double insulated appliances are identified by a symbol comprising a square within a square.

Doublet: A symmetrical antenna fed in the centre with parallel (i.e. balanced) feed line. If ladder or open wire feedline is used a doublet can be connected to a balanced antenna coupling unit to allow operation on multiple bands with only low losses.

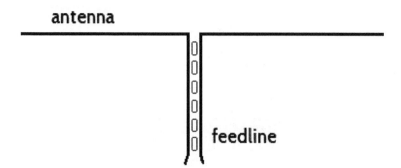

Doubling: See *double*.

Down: Move down in frequency, with the move often specified. For example, "Go down 5" means "go down in frequency by 5 kilohertz". Opposite to *up*.

Downconverter: A radio frequency circuit comprising an oscillator, mixer and filters that converts a higher frequency radio signal to one of lower frequency. May be part of a transverter such as used for microwave communication. Opposite to *upconverter*.

Downlink: The transmission from a satellite that you receive.

Downlink frequency: The frequency you monitor to hear signals from a satellite.

DPDT: Double Pole Double Throw. A type of switch with two SPDT switches controlled by the one actuator. Used where you need to switch two wires at once, for instance the active and neutral leads in a mains power connection.

DR: Dear. (CW abbreviation)

DR OM: Dear Old Man. Used by many non-English speaking operators as a term of respect during a Morse contact. (CW abbreviation)

Drift: Unwanted frequency change in a transmitter or receiver. Can be caused by temperature variations, faulty components or lack of shielding. Severe drift can make equipment almost unusable. Different modes demand different standards for frequency stability, with digital modes followed by SSB being the most critical and AM being relatively uncritical.

Drive: Refers to the level of input signal fed to an amplifier stage. Too low a drive means the amplifier does not deliver its rated power output while excessive drive can distort the signal or damage components.

Driven element: The element to which the feedline is connected in a beam antenna.

Driver: Driver stage. The RF amplifier stage in a transmitter preceding the final amplifier.

Dry cell: A normally non-rechargeable battery such as the common AAA, AA, C or D carbon zinc or alkaline cells. Dry cells are most suited to low current or occasional use applications.

DSB: Double sideband. Double sideband suppressed carrier. An amplitude modulation-based speech transmitting mode featuring upper and lower sidebands but no carrier. Generated in a balanced modulator, DSB is popular amongst those building simple HF voice transmitters because complex sideband suppression circuitry can be avoided.

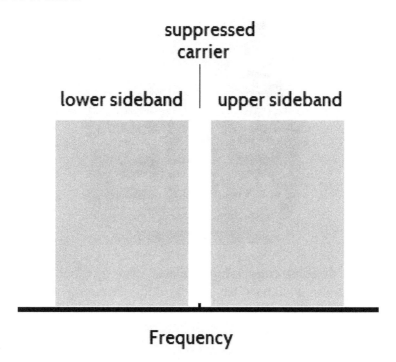

DSP: Digital Signal Processing. An advanced method of analysing and filtering a signal to reduce noise and improve readability. Provided as standard on middle and upper range HF receivers and transceivers.

D-STAR: Digital Smart Technologies for Amateur Radio. A digital voice and data communication mode developed by the

Japan Amateur Radio League and commercialised by Icom. Many cities have VHF and UHF D-STAR repeaters.

DTMF: Dual Tone Multiple Frequency. The telephone touch tone method of sending tones when a keypad is pressed. Used to control repeaters and configure links. See *IRLP*.

DTMF keypad: The keypad often found on a handheld transceiver or external microphones used to send DTMF tones (and perform other functions as well).

Dual (or double) conversion: Relates to a superhet receiver in which incoming signals have two radio frequency conversions (that is two intermediate frequencies) before being converted to audio. This is more complex but is better for VHF and UHF gear as it allows both good image rejection (thanks to the high first IF) and selective IF filtering (easier done at the low second IF).

Dual VFOs: A feature in a transceiver that allows two frequencies to be set up and the user to quickly switch between them. This can be useful for split frequency operating such as often done by DXers.

Dual watch: A feature where a receiver can check two frequencies at once for activity.

Duct tape: A handy accessory in every shack to hold things where nothing else will.

Ducting: Tropospheric ducting. Enhanced VHF and UHF radio propagation, normally caused by an atmospheric temperature inversion, that allows extended range communication over hundreds and sometimes thousands of kilometres. Often occurs where a high-pressure system is nearby. Unlike sporadic-E, ducting is more common on 144 and 432 MHz than 50 MHz.

Dummy load: A high power 50-ohm non-inductive resistor that allows a transmitter to be tested without radiating a signal.

Dupe: Duplicate contact. A contact made in a contest that does not count for points as you have already worked that station.

Duty cycle: The percentage of time that a circuit is on and drawing current. Transmit duty cycle refers to the proportion of time you are transmitting relative to total operating time. This information, which may need to be estimated, is useful when drawing up a power budget such as may be needed when calculating the battery capacity required.

DVM: Digital voltmeter. Test instrument that measures voltage only. Compare with *DMM* which is more versatile.

DX: D = distance, X = unknown. Refers to longer than usual distance amateur communication. Often understood to be outside one's country or continent on most HF bands. On LF, MF, VHF and UHF the distance requirement is relaxed to reflect propagation on these frequencies.

DXAC: DX Advisory Committee. Committee of ARRL that assists in administering the *DXCC* awards program.

DXCC: DX Century Club. An amateur radio award program that gives a certificate to amateurs who obtain confirmation of contact with amateurs in 100 or more countries (or 'entities'). Includes classifications such as mixed or single band and by mode.

DXCC entity: A distinctive geopolitical location that counts for credit towards the *DXCC* award.

DX cluster: An online alert system that advises when sought after stations appear on the air. This often causes pile-ups as hundreds of hams simultaneously come on air.

DX net: An on-air gathering held on a particular frequency and time intended to help stations work rare or long-distance DX stations.

DX station: A station outside your continent or country. Or a station a long way away. See *DX*.

DX window: A narrow segment of band used for DX communication. DX windows may exist through customary usage, specification in a band plan or because the frequencies concerned are shared by countries with differing amateur frequency allocations. Those making local contacts are requested to use other frequencies.

DXer: A radio amateur who works DX, often because they enjoy international communications or are pursuing operating awards.

DXpedition: An organised trip to a remote location, often a rarely active country or island, sought after by *DXers*.

Dynamic microphone: A microphone comprising a diaphragm, moving coil and magnet to generate an electrical impulse from speech. This can then drive an audio amplifier or radio transmitter.

Dynamic range: The range of input signal levels (from very weak to very strong) that an amplifier or receiver can handle without overloading or distorting. Expressed in decibels (dB).

E

E: Electromotive force. Term used to refer to voltage in ohms law formulas. See *EMF*.

E-region: The layer of the ionosphere second closest to earth. Has a maximum concentration approximately 110km above the earth's surface. Can both attenuate and refract signals, in some cases even reflecting them back to earth. The E-region is most known for 'sporadic-E' propagation which allows extended range communication on the higher HF and lower VHF bands, especially over summer. Also see *sporadic E*.

Earth: See *ground*.

Easypal: A popular program, developed by VK4AES, used for transmitting and receiving digital colour pictures. Download from http://www.g0hwc.com/

Echolink: A VoIP-based method of linking radio amateurs. Available on many amateur repeaters, enabling international communication with a handheld transceiver. Registered users who have proved they hold an amateur licence can access Echolink nodes via the internet or an app on their mobile phone. http://www.echolink.org

Eddystone: UK manufacturer of radio equipment. Most known for its HF communication receivers.

EDZ: Extended double zepp. See *double extended zepp*.

EEVBlog: A well-known YouTube channel and discussion forum all about electronics. http://www.eevblog.com

EFHW: End fed half wave. An antenna popular with portable HF operators comprising a half wavelength of wire. Requires a transformer or L-match between it and the transceiver due to its high feed point impedance.

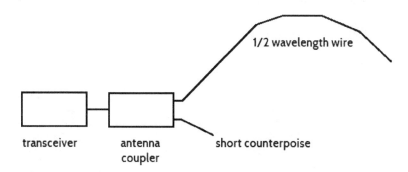

Efficient: Refers to an antenna that radiates almost all power fed to it, with almost none dissipated as heat. A similar concept applies to circuits such as RF power amplifiers where most of the power consumed is converted to RF output on the desired signal rather than dissipated as heat or transmitted on spurious frequencies.

Egg insulator: An insulator made of ceramic or plastic used on the ends of wire antennas.

EHF: Extra high frequency. The range of radio frequencies between 30 and 300 GHz.

Eham.net: A popular amateur website containing reviews of amateur radio equipment and discussion forums. http://www.eham.net

EIRP: Effective Isotropic Radiated Power. Transmitter power multiplied by antenna gain (with respect to a theoretical isotropic radiator). Used to compare the strength of transmitter and antenna installations. Compare with ERP.

Elecraft: US-based manufacturer of transceivers most known for its portable QRP equipment.

ELF: Extremely low frequency. Frequencies between 3 Hz and 30 Hz.

Electret microphone: Small cylindrical microphone that relies on varying capacitance to produce an electrical signal responsive to sound. Normally contains an internal transistor amplifier that needs a small bias voltage to drive.

Electromagnetic spectrum: The range of wavelengths at which electromagnetic radiation takes place, spanning from long wave radio frequencies to above visible light.

Electrolytic capacitor: A polarised capacitor built in a metal can used in applications where high capacitance is required such as in power supplies and audio equipment. Typical values are 1 uF to 10 000 uF or more.

Electron tube: A glass tube in which a current of electrons can flow between electrodes across a vacuum. Used to provide oscillation and amplification before transistors. Also known as a *vacuum tube*.

Electronic log: A log of contacts entered on a computer such as may be submitted as a contest entry. Electronic logs have largely replaced paper logs, at least for larger contests. Data is entered in a computer with logging software used to export it in a standard form that can be uploaded or emailed.

Element: A conductive wire, rod or loop that forms part of an antenna, especially a beam. For example, a yagi's reflector, driven element and directors are all elements. These are normally mounted to or through a central boom which is then fastened to the mast, often via a rotator.

Elevation: Relates to the angle above the horizon. Measured in degrees. Horizon is 0 degrees and straight up is 90 degrees. Along with azimuth this is used to describe the position of a satellite. See *azimuth*.

Elevation angle: The angle formed between the ground and a line between an object in the sky (e.g. a satellite) and the viewer or transmitting station.

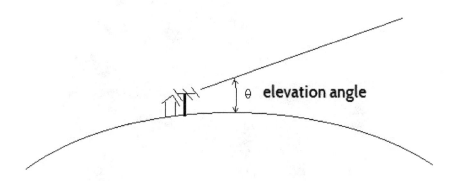

Elmer: A mentor, especially someone who voluntarily coaches newcomers to amateur radio (USA).

Emcomm: Emergency communications. This is an important facet of amateur radio in some countries. (USA)

EMC: Electro-magnetic compatibility. Extent to which electrical and electronic items can operate alongside each other without being the cause or recipient of unintended interference due to electromagnetic energy.

EME (propagation mode): Earth Moon Earth or "moon bounce". VHF or UHF amateur communication based on bouncing signals off the moon. Used to require large antennas and high power but digital modes have made EME possible with modest stations.

EME (energy): Electromagnetic energy. Or Electromagnetic Emissions. See EMR.

EMF: Electromotive force. A potential difference between two points that if bridged to form a circuit causes a current to flow. Measured in volts.

EMI: Electromagnetic interference. Disturbance to the proper operation of electrical or electronic equipment due to poor design or construction that admits or radiates stray electromagnetic signals.

Emission designator: A method of classifying and describing radio transmitting modes using a lettered and numbered code. Emission designators typically include the bandwidth (in Hz, kHz or MHz), the type of modulation (e.g. amplitude or frequency) and the type of intelligence a signal conveys.

Emitter: A connection on a bipolar junction transistor. When wired in a circuit it is normally the connection nearest negative on an NPN transistor and the most positive on a PNP connection. In a common emitter amplifier configuration, the emitter is common to both input and output circuits and may be connected to the earth rail directly or via a current-limiting resistor. In this configuration a small base – emitter current flow (or signal) causes a larger collector – emitter flow (or signal).

EMR: Electromagnetic Radiation. A kind of radiation comprising electric and magnetic fields. Includes radio waves, gamma rays, x-rays and visible light. Also known as electromagnetic energy.

EMR assessment: An assessment done to determine exposure to electromagnetic energy such as from radio transmitters. Required to provide assurance that people are not receiving an unsafe level of exposure. Factors that tend to increase exposure include proximity to radiating antennas, antenna gain, transmitting time, duty cycle and frequency.

Enamelled copper wire: Wire, such as used in motors, relays, transformers and inductors, coated with a thin enamel insulation. This allows close windings without shorted turns. Enamel should be scraped off with a penknife or similar before soldering.

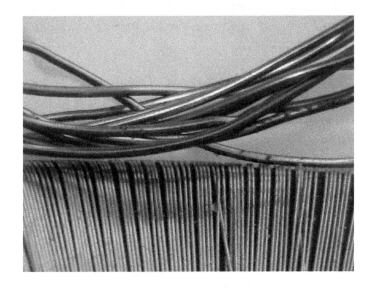

End-fed: Refers to an antenna (normally made of wire) fed at the end. Widely used by portable amateur operators, end-feds avoid the need to take bulky feedline. Half wavelength is a common radiator length. However, the high impedance this length exhibits requires a transformer or antenna coupler between the wire and the transceiver to operate efficiently.

Envelope detector: A type of receiver detector, often comprising a single diode, used to convert AM signals to audio frequencies in simple receivers.

eQSL: A website that permits electronic confirmation of contact such as may be required for awards. Such systems have replaced paper QSL cards for many hams. http://www.eqsl.cc

ERP: Effective Radiated Power. Measured in watts, ERP is the power fed to the antenna multiplied by the antenna's gain relative to a half wavelength dipole. EIRP is similar but uses an isotropic radiator as the reference antenna.

Es: Sporadic E radio propagation. A volatile propagation mode that extends upper HF and VHF communication range. Particularly prevalent over summer on the 28, 50, 70 and (to a lesser extent)

144 MHz bands. Easily recognised by strong signals from stations about 500 to 1500 km away that would not be heard under normal conditions.

ESD: Electrostatic discharge. Sudden flow of electricity between two differently-charged objects due to a path existing between them. Can cause sound or sparks. Electrostatic discharge can damage sensitive components such as ICs. Precautionary steps, such as grounded soldering iron tips and anti-static wrist straps, are often taken to minimise this risk.

ESR: Equivalent Series Resistance. Effective Series Resistance. The effective resistance of a capacitor's equivalent circuit. The lower the better. Electrolytic capacitors are prone to developing high ESR as they age, commonly leading to equipment faults. Measured with a dedicated instrument called an ESR meter.

Etchant: Etching solution. Chemical solution that eats away exposed copper, such as required to make tracks insulated from one another on a printed circuit board. Examples include ferric chloride and ammonium persulphate.

Exchange: Swapping of information between stations to make a radio contact valid. In a contest this often involves the giving and receiving of signal reports and serial numbers which are then logged. An accurate contest exchange is required for a contest contact to be valid.

Exclusive: Refers to a frequency allocation that is allocated for exclusive use by one class of user, normally through ITU agreements.

External antenna: An antenna mounted outside a receiver or room. Allows better sensitivity and freedom from noise. Opposite to *internal antenna*.

External speaker: A speaker plugged in to a receiver or transceiver. Normally bigger and better sounding than the internal speaker.

Extra, Full or Advanced licence: The level of licence that offers full amateur operating privileges. Requires a pass in the most advanced available examination.

Eyeball: A meeting of amateurs in person, not on air.

EZNEC: Popular antenna modelling software. http://www.eznec.com

F

F-layer: The highest region in the ionosphere important for long-range HF communication. Can split into two (known as F1 and F2) during the day. F1's region is typically most concentrated at 300km while F2 has a concentration around 400km. Subject to sufficient ionisation allowing it to form, the F-layer is a reflector of radio signals. Increased ionisation supports propagation at higher and higher frequencies even extending into the lower VHF region during solar peak years. Long-distance propagation via the F-layer is very efficient due to its greater height and lower air density.

Fan dipole: A set of parallel wire dipole antennas fed off the one feedline to allow operation on multiple bands.

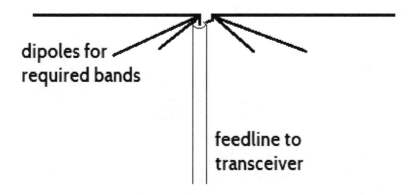

dipoles for required bands

feedline to transceiver

Farad: The basic unit of capacitance. 1 farad is very large and in electronics smaller units such as microfarads (uF), nanofarads (nF) and picofarads (pF) are more widely used.

Farnsworth timing: A method of teaching Morse code where characters are sent at a higher speed but spaces between them are lengthened until the learner gains proficiency.

Fast blow: Refers to a type of fuse that blows almost immediately if an attempt is made to pass current beyond its rating.

Fast charger: A type of battery charger that feeds a high current into a rechargeable battery. Charging time may be a few hours or less. A key risk is heat build-up through overcharging that can damage your battery. Better chargers contain sensors that monitor the battery's state of charge and switch off when charged. Using the correct charger for your type of battery is also very important.

FB: Fine Business. OK or Good. (CW abbreviation)

FCC: Federal Communications Commission. The government radio spectrum regulating agency in the United States. https://www.fcc.gov/

Feedline: The cable that conveys signals between the transceiver (or antenna coupler) and the antenna. Can be either unbalanced (e.g. coaxial cable) or balanced (e.g. open wire feeder). Also *transmission line*.

Feedline loss: The reduction in power level due to inefficiency in a transmission line. Loss tends to increase with frequency, with high-grade cable required for long runs, especially at VHF and above. Often measured in decibels per 100 metres (or 100 feet).

Feedthrough capacitor: A type of capacitor inserted between shielded compartments separating sections of radio frequency circuits to allow DC power to pass while preserving isolation between stages.

Female: Refers to a socket or receptacle into which a (male) plug or pin is inserted to form an electrical connection. While it is most common for female connectors to be on equipment and males to be on cables, this is not always the case. For example, many older CB and ham HF transceivers had a 4 or 8 pin male socket (i.e. containing pins but recessed) into which a female at the end of the microphone's curly cord plugs in to.

Ferrite: An iron powder material used for inductors, broadband RF transformers, baluns and interference suppression. Ferrite has higher permeability than air so when made the core of an inductor it increases inductance. A threaded ferrite, called a slug, can be screwed in or out to allow inductance to be varied.

Ferrite loopstick: A coil of wire wound on a ferrite rod, such as used in AM broadcast receivers as the receiving antenna.

Ferrite rod: A rod made of ferrite as common in AM broadcast receivers. Rods are normally about 8mm diameter and 50 to 200mm long. Alternatively they may be bar-shaped. Amateur applications include use in baluns for HF antennas and interference suppression.

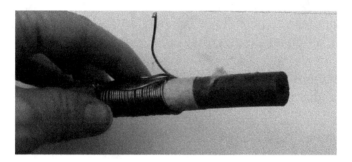

FET: Field Effect Transistor. A type of transistor commonly used in amateur radio transmitter and receiver circuits. Has gate, source and drain connections. Variants include JFET (junction FET) and MOSFET (Metal Oxide Semiconductor FET). Rare types, known as 'dual gate FETs', popular in RF mixer circuits, have two gate connections.

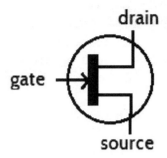

Field day: An organised event where radio amateurs set up stations away from home. This may be to promote amateur radio or as part of a contest.

Field portable: Refers to operating away from one's normal home station address, especially if an outside location without mains power or set-up antennas.

Field strength meter: Meter used to indicate the strength of a nearby radiated RF signal. Useful when adjusting or testing antennas.

Filter: One of the basic stages in radio circuits. Filters can either be low pass (cut off frequencies above a certain value), high pass (cut off frequencies below a certain value), band pass (allow only a narrow band of signals to pass) or band reject (reject a narrow band of frequencies). They are useful to pick off desired frequencies and reject undesired frequencies in transmitters and receivers. Filters can be built for either radio or audio frequencies. Basic passive filters use capacitors and/or inductors while active filters can use transistors or ICs.

Filter capacitor: A capacitor typically used in a power supply to filter noise and ensure that its output is clean. Filter capacitors are mainly high value tantalum or electrolytic type, although lower value disc ceramic types may be connected in parallel to filter higher frequency noise and transients.

Filter method: A method of generating single sideband based on using a narrow radio frequency crystal filter that is so sharp that it can cut one sideband off a double sideband signal. This signal is then converted to the desired transmitting frequency with a mixer and variable frequency oscillator. The reverse can happen on receive, with superhet techniques being used.

Final: Final amplifier. RF power amplifier. The last RF power amplifier stage in a transmitter before the antenna. Uses high power transistors or vacuum tubes.

Final final: Colloquial term for what a station believes will be their last transmission before signing.

Fine tuning: A control on a receiver to slightly adjust its frequency for best reception, such as required when tuning in an SSB, CW or digital mode transmission. Also *RIT*.

Fist: Refers to one's ability to send Morse code.

Fists: An international radio club devoted to the preservation of Morse communication. http://www.fists.org

Flat top: Refers to a dipole or doublet type antenna whose top is perfectly horizontal, i.e. not an inverted-vee.

Flat topping: Describes a signal that sounds distorted (or 'clipped') on its peaks. This effect can be caused by a non-linear amplifier stage or if excessive drive power is applied. When viewed on an oscilloscope the distorted signal has flat tops (like a square wave) even if a signal of varying level is being applied.

Flea power: Colloquial term for very low transmit power. Also *QRPp*.

Flex Radio: A manufacturer of software defined radios. http://www.flexradio.com

Flip Flop: An electronic latch circuit that can have two stable states. Familiar to many as a two-transistor light flasher circuit, flip flops are also a basic building block of digital logic circuits, for example as a means of storing information.

Flutter: Fast variation in the strength of a signal, often heard if either you or the station you are hearing is moving because of an altered signal path (which can even be caused by aircraft flying overhead). Also *mobile flutter*.

FM: Frequency Modulation. A speech mode commonly used for radio communication above 29 MHz. Voice intelligence is conveyed by slightly varying the frequency of a signal. While less efficient than SSB, FM is known for its clear and crisp audio provided signal levels are good. FM is the dominant analogue voice mode on 2m and 70cm where many repeater stations exist to extend transmitting ranges. Amateur satellites such as AO91 and AO92 also support FM.

FMing: Frequency variation of a radio transmitter causing distortion on SSB and chirp on CW. Caused by poor equipment design or voltage drop caused by depleted batteries, an insufficiently sized power supply, long thin DC power leads or excessive transmit power. See *voltage drop*.

FOC: First-class Operators Club. An invitation-based Morse code operators club. http://www.g4foc.org

Folded dipole: A half wave dipole with additional metal added to form a squashed loop with a one wavelength perimeter. Folding quadruples the feed point impedance. This may be useful if using high impedance ribbon feedline or if you wish to compensate for the impedance lowering action of adding parasitic elements when building a beam.

Former: Refers to a non-metallic cylinder on which a coil of wire is wound to form an inductance. Formers may be solid or hollow. Hollow formers may have a slug that is screwed in and out to vary the inductance.

Forward power: Refers to energy that makes its way from the transmitter towards the antenna via the feedline. As opposed to reflected power that is returned when an impedance mismatch is present. Collision between the two produces a 'standing wave', whose level indicates the extent of any mismatch.

Foundation licence: The name of the entry-level amateur radio licence in UK and Australia. Foundation licensees are typically able to use a limited number of modes at low transmit powers on a limited number of bands. There may also be restrictions on homebrew equipment.

Fox: The hidden transmitter used in a foxhunt, T-hunt or *ARDF* event.

Foxhunt: A hide-and-seek radio sport involving the use of radio direction finding to find one or more hidden transmitters. Does not involve animals. See *ARDF*.

Free DV: A mode for digital voice communication. http://www.freedv.org

Free-running: Free-running VFO. Refers to a radio frequency oscillator whose frequency is governed by an LC tuned circuit (and not a crystal or signal derived from one such as a frequency synthesiser). Simple to build but often unstable, free-running

VFOs have been replaced by PLL or DDS frequency synthesisers in modern equipment.

Freq: Frequency. (CW abbreviation)

Frequency: The number of occurrences per unit of time of an event. In radio this is expressed in cycles per second or hertz, with kHz, MHz and GHz being multiples of one thousand, one million and thousand million respectively. Frequency is inversely proportional to wavelength – i.e. long wave signals are low frequency while short wave signals are high frequency.

Frequency agile: Refers to a receiver or transmitter that can be set to any frequency in a band due to having a free-running oscillator or frequency synthesiser. Opposite to *crystal controlled.*

Frequency coordination: An activity done by or for beacon and repeater sponsors to ensure that operating frequencies are chosen to not interfere with other communication facilities or spectrum users.

Frequency counter: An instrument that measures and displays frequency. Useful for testing radio frequency oscillators and transmitters.

Frequency coverage: The range of frequencies that a receiver, transmitter or transceiver can cover. Also *tuning range.*

Frequency divider: A circuit that divides an incoming frequency to an integer submultiple, e.g. one half, one third, one quarter, one tenth etc. Used in both analogue and digital circuit applications.

Frequency hopping: A way of transmitting radio signals that involves rapid jumps in frequency. Favoured by military and other users for whom security against interception and interference is important.

Frequency multiplier: A stage in a radio circuit that multiplies the incoming frequency by two, three or more times. May be passive, involving diodes, or active, involving transistors. Frequency multipliers were commonly used in the early days of radio to obtain a higher frequency from a lower frequency crystal. Technique continues to be used today for UHF and microwave communication.

Frequency reference: An oscillator of known stable and accurate frequency used to ensure that other equipment is also indicating and operating on the correct frequency.

Frequency response: The range of frequencies that an amplifier, microphone or speaker can efficiently operate over. The term is most often used for audio equipment but a similar concept applies for RF amplifiers, filters, antennas, etc.

Frequency synthesiser: Circuitry that generates a radio frequency signal derived from a crystal-controlled reference. Can normally be varied in frequency with an LCD digital frequency display. See *DDS*.

Front end: The first few stages of a radio receiver. Typically include band pass filter, RF amplifier and first mixer stages.

Front-to-back ratio: The extent to which the strength of the signal radiated off the front of a directional antenna exceeds that of the signal radiated from its back. Longer and more directive beams have higher front to back ratios. Measured in decibels down on the main (front) lobe. As an example, the antenna below has a 25dB front-to-back ratio.

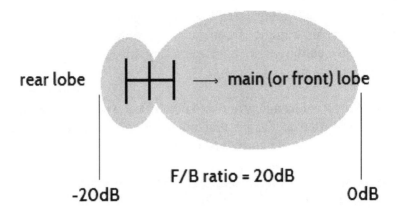

Front-to-side ratio: The extent to which the strength of the signal radiated off the front of a directional antenna exceeds that of the signal radiated from its side. Longer and more directive beams have higher front to side ratios. Measured in decibels down on the main (front) lobe.

FSK: Frequency shift keying. A mode of transmission based on small frequency changes to a carrier signal to convey intelligence. Used for Morse code keying on some beacons, radio teletype and some digital modes.

FSTV: Fast scan television. Regular television involving moving pictures and a wide transmission bandwidth. Distinct from SSTV, comprising still pictures and a narrow bandwidth, and NBTV, comprising moving but low-resolution pictures transmitted in a narrow bandwidth.

fT: The gain bandwidth product of a transistor. Normally expressed in MHz, this is the maximum frequency that a transistor will amplifier a signal at. You should use a transistor with an fT several times higher than the maximum frequency your circuit operates at to assure sufficient gain.

FT: Letters that precede the model number of most Yaesu transmitting equipment.

FT8: A partly-automated digital mode known to be very effective under weak signal conditions. Requiring accurate timing, transmissions take place in 15 second cycles. FT8 allows exchanges but not conversations.

FT8 Call: A version of the FT8 digital mode that allows keyboard to keyboard conversations. Now called *JS8 Call*.

FT243: The case or holder style of an old-fashioned quartz crystal. Identified by their bakelite case with screws that allow the crystal to be opened. FT243 holders have thick pins spaced to allow insertion into an octal tube socket.

Full duplex: Amateur communication that allows you to listen while you are talking, similar to speaking on a telephone. Typically requires equipment on different bands and you wearing headphones to prevent audio feedback. See *cross-band*.

Full wave loop: A wire loop antenna with a perimeter of a full wavelength. May either be mounted vertically, for example as the driven element in a cubical quad, or horizontally, with its plane parallel to the ground.

Full wave rectifier: a network of diodes that conducts during all phases of an AC's current's cycle. Full wave rectifiers are a popular and efficient choice in AC to DC power supply circuits.

Fully quieting: Refers to a signal with a carrier strong enough to silence the noise generated in an FM receiver.

Fundamental: The intended frequency of a signal from an oscillator or transmitter. Compare to *spurii* or *harmonic*.

Fundamental overload: Interference caused by the device whose operation is impaired being susceptible to strong signals such as from a nearby transmitter. The fault lies with the equipment being interfered with rather than with the transmitter that may be accused of causing the interference. Treatments include improved shielding, grounding and filtering to lessen the entry of RF energy.

Fuse: A glass or ceramic component consisting of a thin wire that burns out if excessive current is drawn, such as may happen if an item is connected reverse polarity or a fault develops. Its purpose is to protect equipment from further damage.

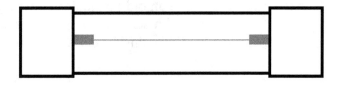

Fusion: See *System Fusion*.

G

G-QRP Club: A UK-based club devoted to low power (QRP) communication. http://www.gqrp.com

Gain: Refers to the amplification ratio of an electrical circuit or antenna. Gain is typically measured in decibels to allow easy comparisons. A high gain amplifier, receiver or antenna can be expected to provide a stronger signal than a low gain amplifier, receiver or antenna.

Gain distribution: Refers to the locations within equipment where signal amplification and attenuation occurs. For example, a superhet receiver may have most of its gain in its RF front end and IF amplifier stages while a simple direct conversion receiver may have most of its amplification at audio frequencies. Some stages, such as filters, attenuate signals so need a preceding or following amplifier. Inappropriate gain distribution in a receiver can lead to a set that is insensitive, has too much internal noise or is poor at handling strong signals.

Gamma match: A method of transforming the impedance of a feedline to that presented by the driven element of yagi or quad antenna. Uses a short metal arm running from the centre and parallel to the driven element for a short distance and a small capacitor to tune out reactance.

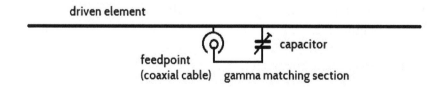

GaAsFET: Gallium arsenide field effect transistor. A type of field effect transistor used in VHF and UHF radio circuitry.

GDO: Grid dip oscillator. An old name for a dip oscillator useful for various tuned circuit and antenna checks.

General, standard or intermediate license: The middle level of amateur radio licence in some countries. Requires a pass in an intermediate level exam. Provides a comprehensive selection of amateur privileges but not necessarily all frequencies or power levels.

General coverage: Refers to radio equipment that can receive or transmit on all HF bands, not just amateur frequencies.

Gentlemen's agreement: Convention that amateurs are 'gentlemen' and 'never knowingly use the airwaves in a manner that lessens the enjoyment of others'. More specifically relates to operating procedures and band plans that are voluntary and not mandated in radio regulations. *See Amateurs Code* and *self-regulation.*

Geostationary: Refers to a satellite that, due to its special position and altitude (36 000 km over the equator), occupies a fixed position relative to earth. This permits all-day use of them and dishes to be pointed in a fixed position.

Germanium diode: A type of sensitive diode commonly used in crystal set receivers and radio test equipment.

Gimmick capacitor: A low value trimmer capacitor made by twisting two pieces of insulated wire together. The resultant 1 to 5 pF capacitance is used in certain applications such as coupling VHF signals between stages or neutralising RF power amplifier stages. Tightening or loosening the twist varies the capacitance.

GMT: Greenwich Mean Time. See UTC.

Gnd: Ground. Ground connection. See *Ground.*

Go box (or bag): A container containing items necessary to put a field portable amateur radio station on the air. It may also include items such as food and drink required to sustain time away from home.

Golden screwdriver brigade: The self-regarded expert who does more harm than good to their transmitter, often in a mistaken quest to increase its output power. Also *screwdriver expert.*

GPS: Global Positioning System. A satellite-based system used for navigating, position finding or accurate time-keeping in conjunction with portable receivers.

Great circle: The shortest distance between two points on the globe. Knowing the direction of this from your location can help you point your beam in the right direction. Also see *Azimuthal map.*

Green stamp: Term for a $US 1 note. These are sometimes sent in the mail when you would like a contacted station to send a QSL card directly to you and wish to pay their postage. This is most likely to be the case when the other station is a DXpedition or is rarer than you.

Greyline: Refers to the period around sunrise and sunset where long distance HF radio propagation is enhanced and DX is more easily worked. This enhanced propagation is partly due to the absence of the ionosphere's lossy D-layer, which disappears quickly at sunset but takes some time to appear after sunrise.

Grid: A connection on a vacuum tube. Triodes have one grid (control grid), tetrodes have two grids (control grid, screen grid), pentodes have three grids (control grid, screen grid, suppressor grid). The control grid controls the flow of electrons from the cathode to the anode and is often (but not always) the connection to which input signals are applied.

Grid locator system: A system of dividing the word up into 'squares' based on latitude and longitude. Each square has a 4-character code, with the option of longer codes for more exact placement. Grid locators are useful to convey location information in few characters such as handy for some digital modes. They are also the basis of some contests and awards that score contestants on the number of squares worked.

Grid square: A geographical division of the world based on latitude and longitude under the *grid locator system*. The 'squares' are actually more like trapeziums, with the top and bottom side lengths becoming increasingly unequal towards the poles.

Grinding: Refers to an old practice where quartz crystals could be opened up and ground to change their frequency slightly. This was common in the early days of SSB to make crystal filters that required crystals to be on slightly different frequencies.

Ground (electrical): Refers to an electrical connection to earth. May be for various purposes, including as part of an electrical distribution system, diversion of lightning-induced static electricity, anti-static protection of components, and for an antenna system.

Ground (electronic circuit): The reference point in an electrical circuit against which voltages are measured for adjustment and troubleshooting purposes. It may be labelled with the earth symbol or 0 volts. In many simple devices ground is often, but not always, the negative power supply connection. More complex gear, especially that using op-amps, may have both positive and negative supply rails, with voltage measurements being done against ground. When looking at a circuit diagram ground can often be recognised by the large number of capacitors, resistors, inputs and outputs with one side connected to it. A circuit's ground may be connected to an item's metal case (where provided) and, if it plugs in to the wall, the power plug's earth pin.

Grounding: Refers to the contact that equipment has to ground, especially where required for safety or other purposes.

Ground plane (antenna): A type of vertical antenna comprising a quarter wavelength upright element and between two and four horizontal or sloping down radials. Rigid designs using tubular aluminium are common on VHF and lower UHF while wire ground planes are popular HF antenna projects. The coaxial feedline's inner connects to the vertical element while the braid connects to the radials.

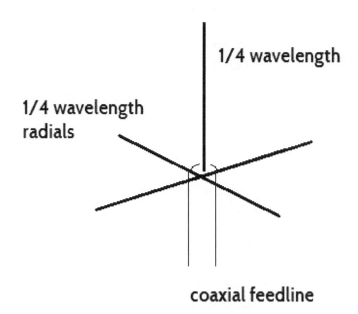

1/4 wavelength

1/4 wavelength radials

coaxial feedline

Ground plane (circuit board): A large area of copper left unetched on a printed circuit board. Normally tied to power supply ground, its purpose is to provide a convenient low-loss connection point for noise and frequency-sensitive components in critical radio, audio and digital circuits.

Ground rod: A copper rod hammered into the ground to provide a contact with earth as used for electrical, station or antenna ground radial installations. Also known as ground stake or earth stake.

Ground wave: Refers to radio propagation that follows the curvature of the earth. Ground wave range is longest at low and medium frequencies, where it is the dominant propagation mode. Contrasts with sky wave propagation that is the main propagation mode at high frequencies and direct wave on VHF/UHF.

Grub screw: A small headless screw used to fasten knobs to some types of rotary switch, potentiometer or variable capacitor shafts.

GUD: Good. (CW abbreviation)

Gunn diode: A rare type of diode that could operate at microwave frequencies. Commonly used in early solid state SHF equipment, gunn diodes could be made to oscillate due to their 'negative resistance' characteristics.

Guy wire: A wire support for a tower or mast to keep it stable in high winds.

G5RV: A popular horizontal wire antenna used on HF. A form of centre fed doublet 31 metres (102 feet) end to end. Approximately 10 metres of slotted ribbon is used as the feedline. This goes either to a balanced antenna coupler or 1:1 balun to which coaxial cable is connected. Efficient operation is possible on most bands between 80 and 10 metres.

H

H: Henry. The unit of inductance. 1 henry is very large and in radio smaller units such as millihenry (mH) and microhenry (uH) and nanohenry (nH) are more widely used.

HAAT: Height above average terrain. How high a transmitting point is above surrounding terrain.

Hairpin match: Loop of wire on a beam antenna's driven element used to provide an impedance transformation for the feedline.

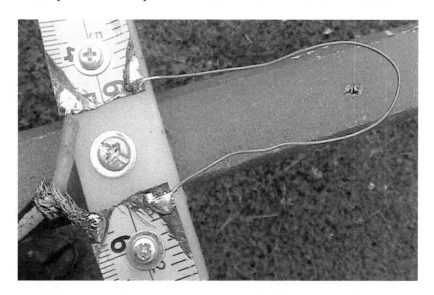

Half duplex: Refers to a radio communications circuit where you can either talk or listen but not both simultaneously (like you can with a telephone which is full duplex).

Half-lattice: A style of crystal filter, typically using crystals slightly separated from one another in frequency, common in superhet receivers and SSB transceivers.

Half-sloper: A wire antenna comprising a sloping quarter wavelength of wire, most often suspended from a metal tower which may be used to support other antennas. The half-sloper's wire is insulated from both tower and ground with the top being connected to the centre conductor of its coaxial feedline. The feedline's braid connects to the tower. Half-slopers are favoured on the 1.8 and 3.5 MHz bands as they don't need as tall a tower as a full quarter wavelength vertical would. Half-slopers are conveniently unobtrusive as they can double as one of the mast's guy wires.

Half square: A type of vertically polarised wire antenna used on HF. Contains one wavelength of wire bent into a shallow inverted-u shape. Feeding can either be in one of the top corners (low impedance) or at the bottom of one of the legs (high impedance).

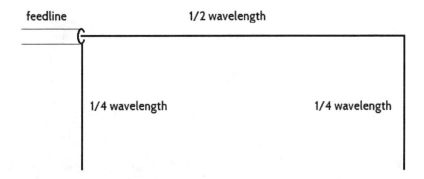

Half wavelength dipole: A popular antenna comprising a half wavelength of wire fed in the middle. In free space it has a bidirectional figure of 8 shaped radiation pattern with maximum signal broadside to the wire. Often used as the standard against which other antennas are compared. A dipole may be fed directly with 50 to 75 ohm coaxial cable or via a 1:1 balun. Or, if coverage of multiple bands is required, use could be made of high impedance open wire or ladder line with a balanced antenna coupler.

Half wavelength end-fed: A half wavelength of wire fed at the end to form an antenna. Its high feed point impedance (several thousand ohm) means it must be connected to the transceiver via a transformer or antenna coupler to provide the required impedance transformation to 50 ohm. A short counterpoise is also often used. Popular with portable operators as it is easy to erect and requires no heavy feedline. Also see *EFHW*.

Halo: A type of horizontally polarised omnidirectional antenna sometimes used by VHF SSB mobile operators.

Halyard: Rope used to raise and support a wire antenna up a pole or mast.

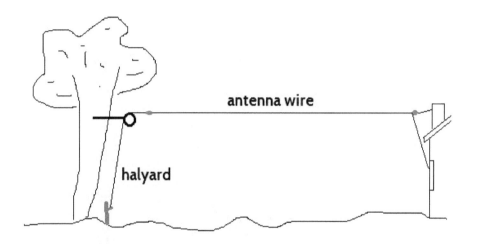

Ham: A licensed radio amateur. Or relating to amateur radio. Never 'HAM'.

Hamfest: A flea market-style gathering where hams buy and sell radio and electronic equipment. There may also be club displays, competitions and technical forums. Typically organised by radio clubs. Potentially hazardous to your wallet.

Ham radio: Amateur radio.

Ham spirit: Camaraderie and goodwill between radio amateurs similar to the Amateurs Code. Includes assisting and mentoring newcomers, answering other people's CQ calls, generosity with information, fostering international friendship and hospitality, not interfering with others, etc.

Hand capacitance: A variation in a receiver's frequency caused by moving your hand near the front panel and its tuning capacitor. Caused by poor earthing or shielding, hand-capacitance was common in homebrew receivers with free-running oscillators built in wood or plastic cases.

Handle: a person's first name. Term is disliked by some and it's mostly easier just to say 'name'.

Hand key: A manual up and down key for sending Morse code.

Harmonic (radio signal): A multiple of a signal's frequency. Can either be desired (as with a frequency multiplier) or undesired (as in a transmitter with spurious outputs). The latter can cause interference, especially if the multiple is outside an amateur band. Correct transmitter adjustment and/or a low pass filter can minimise the generation of harmonics.

Harmonic (person): Amateur slang for a child, especially one's own.

Harmonic suppression: Refers to the extent to which a transmitter, RF power amplifier or RF filter suppresses harmonics.

Equipment with poor harmonic suppression can cause interference and is below mandated technical standards. Fortunately, an external low pass filter can improve substandard harmonic suppression.

Harmonically related: Refers to bands whose frequencies are multiples of one another. For example, the amateur 1.8, 3.5, 7, 14, 21 and 28 MHz bands. Or 144, 432 and 1296 MHz. As well as simplifying the design of transmitters with frequency multipliers (as common in the 1930s – 1950s), this arrangement ensured that spurious harmonics amateurs generated mostly only interfered with themselves and not other radio spectrum users.

Hash: RF interference such as commonly found in urban neighbourhoods due to noise from power lines, switch mode power supplies, solar systems, LED lighting etc. Can sometimes be reduced by shielding equipment or fitting RF-blocking ferrites on external leads.

HB9CV: A type of compact two element beam antenna. Often used for VHF foxhunting and direction finding due to its close element spacing. Named after Swiss radio amateur HB9CV.

HC: Letters often used precede the holder style of quartz crystals (e.g. HC6, HC18, HC25, HC49 etc). These are metal cases not designed to be opened (although they can).

Heathkit: US manufacturer of electronic and amateur radio kits most active in the 1950s to 1980s.

Heatshrink tubing: Plastic tubing that shrinks when heat is applied. Useful for insulating electrical connections.

Heat sink: A piece of metal attached to a component such as a power transistor to dissipate heat away from it to prevent overheating.

Heat sink compound: A white paste applied between a component and its heat sink to aid thermal conductivity and thus the ability of the heat sink to dissipate heat.

Helical: Refers to an antenna formed from a coil of wire. Often used where space is limited such as on hand held equipment or vehicles. Saves space compared to a full-sized antenna but less efficient.

Helical resonator: A type of sharp bandpass filter used for VHF and UHF filtering in receivers, transmitters, etc.

Hepburn index: A scale, developed by William Hepburn, used to forecast and indicate VHF/UHF enhanced propagation due to tropospheric ducting on a map. http://dxinfocentre.com

Hertz antenna: A non-grounded half wave dipole antenna. Named after its inventor.

Heterodyne: Refers to a beat signal or note generated from the mixing of two signals. This can be desirable (in the case of receiver product detector stages) or undesirable in the case of interference to reception.

Hex beam: A type of compact wire beam antenna supported on a hexagonal structure typically made from lightweight fibre or bamboo poles. Provide low-cost gain and directivity on the upper HF bands.

hFE: Hybrid parameter forward current gain common emitter. A measure of a transistor's DC gain. Many multimeters indicate hFE on their transistor test function.

Hi: An expression of laughter. (CW abbreviation)

Hi-Hi: An affected term sometimes used to indicate laughter when communicating by voice.

Hidden transmitter problem: An issue faced with automated digital modes such as packet radio where not all users' systems can hear one another. This causes stations to unwittingly transmit over one another and slows data transfer.

High angle radiator: Refers to an HF transmitting antenna that sends most of its energy straight up or at angles close to 90 degrees from the earth's surface. This is useful for communication up to about 500km in the 1.8 – 7 MHz frequency range but poor if DX contacts are desired. Typical examples include low horizontal antennas such as dipoles and loops erected with their plane parallel to the ground. Opposite to *low angle radiator*.

High gain: Refers to an amplifier with a high amplification ratio. Or an antenna with substantially more gain than a reference antenna such as a half wave dipole or the theoretical omnidirectional isotropic radiator. Achieves gain by concentrating (or 'beaming') the signal in one direction through the use of parasitic elements or, at microwave frequencies, a dish reflector.

High-Z: High impedance. Refers to signals that have relatively high voltages but low currents for a given power level. Often used with reference to headphones (for crystal sets), microphones or antennas.

HOA: Home Owners Association. An association of home owners that administers common property in certain types of residential development. May have regulations with regards to the erection of antennas that restrict amateur activities (USA)

Hollow state: A playful term used to refer to electronic equipment that uses vacuum tubes. Opposite of *solid state*.

Homebrew: Refers to equipment that amateurs have assembled from a circuit diagram. There is a special satisfaction in receiving signals and making contacts with such equipment.

Homebrewer: A radio amateur who builds their own equipment. Because many builders are on to the next project almost as soon as their last is complete, there are more around than casual listening might suggest.

Honour Roll: A prestigious division of the DXCC Awards program open to those who have worked station in the most entities.

Horizontally polarised: Refers to an antenna that emits or efficiently responds to horizontally polarised signals, i.e. those whose lines of electric flux are in the horizontal plane.

Horn: Directional horn-shaped antenna used on microwave frequencies.

Hot carrier diode: A type of diode used for high speed switching, as useful in RF circuits. Also *Schottky diode*.

HP CU AGN: Hope to see you again. (CW abbreviation)

HPF: High pass filter. A filter designed to pass signals above a certain frequency but reject signals below that frequency. An example use is if you live so close to a 1 MHz AM broadcast station that it overloads your amateur receiver on 3.5 MHz. A 3 MHz high pass filter in the antenna line would allow the desired 3.5 MHz signal to pass unaffected but suppress the undesired 1 MHz signal.

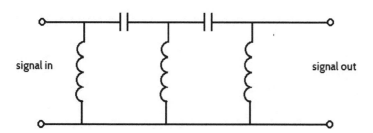

HR: Here. (CW abbreviation)

HRD: Ham Radio Deluxe. A popular amateur radio computer software package incorporating functions such as logging, transceiver control, rotator control, digital modes and satellite tracking. http://www.hamradiodeluxe.com

HT (transceiver): Handie-talkie. Refers to a hand-held battery-powered transceiver. Typically operates on VHF/UHF bands. HTs transmit using FM and/or a digital voice mode. Unless used on hilltops their range is limited so their usefulness is heavily dependent on repeaters.

HT (voltage): High tension. Refers to the high voltages required to operate tube (valve) equipment such as receivers, transmitters and linear amplifiers. Building and repairing such gear requires safety precautions to prevent electric shock.

Hum: Low noise typically derived from an AC mains supply if it is badly filtered. Can interfere with sensitive audio equipment (e.g. stereo amplifiers) or come through on transmissions.

HW: How. (CW abbreviation)

HW?: How are you receiving my signal? (CW abbreviation)

Hy-Gain: US manufacturer of antennas. Well known for its verticals and beams. http://www.hy-gain.com

Hz: Hertz. Unit of frequency. 1 Hz = 1 cycle per second. Multiples include kHz, MHz and GHz.

I

I (current): Term used to refer to current in ohms law formulas.

I (signal): In phase signal. Term often used when explaining phasing SSB or software defined radio systems. Opposite to *Q (signal)*.

Iambic keyer: Morse key alternative. Instead of pressing a button, with you manually timing the length of dits and dahs, a keyer uses a paddle with a side to side motion. Keyer timing circuitry automatically gives a uniform length of characters. See *paddle*.

IARP: International Amateur Radio Permit. Means of allowing visiting amateurs to operate outside their own country. Widely accepted in North and South America. http://www.arrl.org/iarp

IARU: International Amateur Radio Union. The international grouping of national representative societies. Allows representation of amateur radio at an international level. http://www.iaru.org

IBP: International Beacon Project. A system of HF beacons around the world occupying common frequencies. Beacon transmissions are carefully timed to avoid any two transmitting at once. IBP beacons operate on 14.100, 18.110, 21.150, 24.930 and 28.200 MHz. http://www.ncdxf.org/beacon

IC (component): Integrated circuit. A wafer (normally silicon) onto which multiple electronic components have been etched, allowing it to do the work of multiple discrete circuit stages, saving space and power. Main families are analogue (such as audio amplifiers) and digital (such as used in computer logic circuits). Also *silicon chip*.

IC (transceiver model): Letters that precede the model numbers of Icom transceivers.

Icom: A popular Japanese manufacturer of amateur radio equipment. http://www.icom.co.jp/world

Ident: A voice, morse or data part of a signal that identifies a beacon or repeater.

Identify: To send your call sign. Regulations normally specify that you identify at regular intervals when transmitting.

IF: Intermediate frequency. The radio frequency to which incoming signals are converted before being demodulated in a superhet receiver. More complex multi-conversion receivers have multiple intermediate frequencies. Filtering at the final IF normally provides the receiver with its main selectivity.

IF filter: A narrow bandwidth filter that forms the main selectivity determining element of a superhet receiver. It may be crystal, mechanical or, in basic receivers, two or three LC tuned circuits. For best selectivity its bandwidth needs to match the mode being received.

IF gain: Intermediate frequency gain control. A setting found on more advanced receivers that allows the amplification factor (i.e. gain) of a superhet's intermediate frequency amplifier stages to be varied. This may be desirable to obtain gain distribution that optimises performance, especially when strong signals are present.

IF rejection: The ability of a receiver to reject extraneous signals that may appear at one of its intermediate frequencies. High IF rejection requires design and circuitry that provides good shielding, isolation and filtering.

IF shift: A control on a receiver that shifts its intermediate frequency slightly to allow a slightly different lower and upper frequency limit for the IF filter. While rarely needed, such an adjustment may be useful to dodge interference.

ILLW: International Lighthouse and Lightship weekend. An annual event where amateurs operate from lighthouses and lightships around the world. http://www.illw.net

Image: Refers to a signal on a frequency different to that indicated on a superhet receiver's dial or display that is being heard due to poor front-end RF filtering in the receiver. The main image frequency is typically twice the IF away from the indicated 'dial frequency'. Other images can be caused by the local oscillator having spurious outputs and these mixing with strong incoming signals. Poor image rejection is common on cheap receivers with low intermediate frequencies and/or poor front-end filtering.

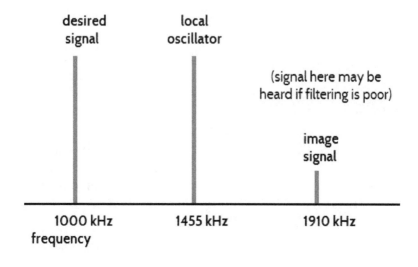

Image rejection: The ability of a receiver to reject signals on image frequencies. A receiver with poor image rejection hears signals on extraneous frequencies that wouldn't be there on a good receiver. Such stray signals can spoil reception of desired signals. Good front-end filtering, clean local oscillator outputs and a well-chosen IF all maximise image rejection.

IMD: Intermodulation distortion. Distortion caused by interaction between multiple amplitude modulated signals in a non-linear system such as a defective or over-driven transmitter or amplifier.

Impedance: In relation to AC signals the ratio of voltage to current. High impedance signals have a high ratio of voltage to current while low impedance signals have a low ratio of current to voltage. A simplified example is two 120 watt light globes of equal brightness. One is designed for 120 volts while another is for 12 volts. At their specified voltages the first draws 1 amp while the second consumes 10 amps. Connecting the 120 volt globe to a 12 volt supply will not light it properly and some means of impedance conversion (otherwise known as a transformer) will be required to deliver 120 volts and achieve full brightness. A similar concept applies with radio transmitters and antennas except some antennas present reactive loads that need to be tuned out for efficient power transfer.

Impedance mismatch: A situation where a load (such as an antenna) exhibits a different impedance to a source (such as a transmitter). Serious mismatches can result in (at best) inefficient power transfer and (at worst) potential equipment damage. Some form of impedance transformation is necessary to transform the resistive component (and if necessary) tune out the reactive portion of the mismatch to provide efficient operation. It's important to note that while transforming impedances in the shack can keep the transmitter happy, significant losses may still sometimes remain, especially if trying to use an antenna away from its design band and loss-prone coaxial feedline is being used.

Impulse interference: Clicking type RF noise as may be generated by motor vehicles, electric motors or electrical equipment being switched on or off.

In phase: Refers to signals (of the same frequency) whose waveforms are in step with one another. Opposite to *out of phase*. Also see *phase difference*.

Incentive licensing: An amateur licensing system with multiple levels. Passing a more advanced exam gives increased frequency and output power operating privileges. Beginners start with a simple entry-level licence. Then, if desired, they can sit a test to advance (or 'upgrade') to one of the higher levels when their skills develop. Most countries have two or three level incentive-type licensing systems but the term is most used to refer to the controversial five-level 1960s US system that required existing amateurs to sit additional tests to retain privileges.

Inductance: The property where an electrical conductor, such as a wire, forces a voltage (or electromotive force) to be generated when the current through the wire changes. This characteristic is useful in many electronic and radio circuit applications.

Inductive coupling: Refers to interaction between two closely spaced (and usually parallel) conductors in a circuit where a change of current in one side causes a change in the other side due to mutual coupling. In some cases, this is unwanted (e.g. causing a radio frequency amplifier to oscillate) while in other cases it is a design feature (such as in transformers and band pass filters).

Inductive reactance: Resistance to AC signals (including those at radio frequencies) caused by inductance (either deliberate or stray) in a circuit. Unlike resistance, whose value is independent of frequency, the reactance of a given inductor increases with frequency. Another property of inductive reactance is that its effect can be cancelled out by capacitance, but only for signals of a particular frequency. This frequency dependent behaviour is what makes tuned circuits work.

Inductor: An electrical component comprising a coil of wire, either wound on a cylinder, passed through a toroid or etched on a circuit board. An inductor has been deliberately made to possess inductance which, like capacitance, exhibits reactance, a special form of resistance that varies with frequency. For example, a perfect series inductor won't affect the flow of a DC current but will retard an AC signal, with its attenuation effect (or reactance) increasing with frequency. Inductors are widely used in electronic circuits. They are commonly teamed up with capacitors to perform many filtering, signal selection and interference rejection functions in receivers and transmitters.

Inner: The centre conductor of coaxial cable such as used for carrying audio and RF signals.

Input frequency: The frequency that a repeater station receives on. This is your transmitting frequency. Use of a repeater requires (i) you to be on its input frequency, (ii) your signal to be sufficiently strong at the repeater site, and (iii) you sending any required subtone the repeater requires.

Instability: Refers to an unwanted change in conditions in a circuit. For example, frequency drift in an oscillator stage or parasitic oscillation in an amplifier stage.

Insulator: Material that has a very high resistance to electricity (e.g. glass or plastic).

Intercept point: Third order intercept point. A measure of the linearity of circuits such as radio receivers when assessing their dynamic range or ability to handle strong signals.

Interface box: A unit that connects between a computer and transceiver to allow operation on digital or sound card-based modes such as PSK31, WSPR, SSTV, FT8 etc. Simpler types connect between the transceiver's microphone and speaker connections and the computer's audio in and out sockets. The box may contain audio isolation transformers, level controls and possibly transmit/receive switching.

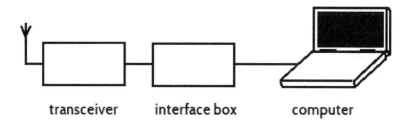

transceiver   interface box   computer

Internal antenna: An antenna inside a receiver, such as a ferrite rod used for MF and LF broadcast reception.

Internal noise: RF noise generated by circuitry within a receiver. Needs to be over-ridden by noise from the antenna if the receiver is to be sufficiently sensitive.

Internal resistance: A characteristic of batteries. Batteries are imagined as comprising cells in series with a resistor. The resistor is harmless for low current applications but leads to voltage drop if heavy currents are drawn. Different types of battery have different levels of internal resistance. Alkaline batteries, with high internal resistance, are suitable for low-drain applications while lead acid, nickel cadmium or lithium-based batteries can drive heavier loads due to their lower internal resistance.

International Morse code: The now universal system of Morse code, comprising sequences of short and long characters (dits and dahs) to form letters, numbers and punctuation symbols as used to send telegraphic messages.

Intruder: A non-amateur station that transmits inside an exclusive amateur band against ITU agreements. They may be a pirate or be operated by foreign governments that do not respect international spectrum conventions.

Inverted L: A type of end-fed wire antenna that has part of it vertical and another part horizontal, tied off to another high support. A good general-purpose antenna for the lower HF bands, it transmits with a mixture of vertical and horizontal polarisation.

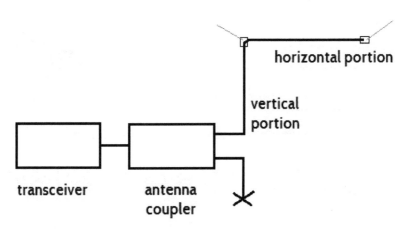

Inverted Vee: A wire dipole antenna that has its ends mounted at lower height than its centre. This reduces its feed point impedance to 50 ohm, which makes it a good match to standard coaxial cable. Another more practical advantage of the inverted vee is the need for only one high support.

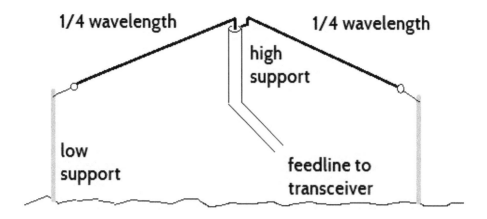

Inverter (logic): A digital circuit that has as its output a logical state opposite to that presented at its input. For example, 0 becomes 1 and 1 becomes 0.

Inverter (power supply): A circuit, using an oscillator and transformer, to convert DC to AC current and also often a different voltage. A common application is converting 12 volts DC from a solar system to the higher AC voltage required to power home appliances.

I/O: Input/Output connections. (computer terminology)

Ionosphere: A series of layers in the earth's upper atmosphere that become ionised when exposed to solar radiation. Located between about 100 and 400 km above the earth's surface, the ionosphere is responsible for absorbing or reflecting radio signals. It is the latter that permits long-distance propagation of MF, HF and lower VHF radio signals. The diagram below shows the ionosphere's main layers, with the F1 and F2 layers combining at night.

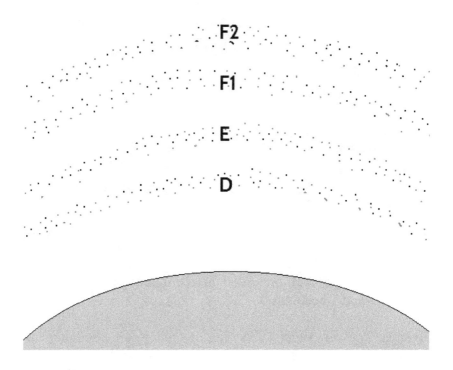

Ionospheric absorption: A property of parts of the ionosphere (notably D layer) to absorb and thus weaken HF radio signals. Absorption typically reaches a maximum in the middle of the day and is the reason why lower HF frequencies are typically dead at this time. Conversely early mornings and evenings, where absorption is least, offer opportunities for long distance communication.

IOTA: Islands On The Air. An awards program that encourages amateurs to make contacts from and with islands around the world. http://www.iota-world.org

IPO: Intercept Point Optimisation. Pretentious name given by Yaesu for a switch that disables a receiver's RF preamplifier, such as desirable to stop very strong signals from overloading a receiver.

IR: Infra-red. Refers to the portion of the electromagnetic spectrum with a frequency above that of the radio spectrum but below that of visible light. IR light, typically emitted by a special IR LED, is used for applications such as short-range remote control of electronic equipment.

IRC: International Reply Coupon. A largely obsolete way of paying for return postage such as if you would like a direct QSL card from someone and wish to pay their postage. Was able to be redeemed at post offices for sufficient stamps to send one international letter.

IRLP: Internet Repeater Linking Project. A method of linking VHF and UHF repeaters by the internet, so allowing international communication with a hand-held transceiver. The DTMF keypad on your transceiver is used to establish a link with any IRLP repeater. http://www.irlp.net

Iron powder: A type of ferromagnetic material onto which wire is wound to form inductors and baluns. May be ring (toroidal) or straight (solenoid) shape.

ISB: Independent sideband. Refers to a double sideband transmission where the sidebands contain different audio content. Rarely used on the amateur bands but could be used to enable crossband communication or allow a simultaneous slow scan TV and voice transmission.

ISM: Industrial, Scientific, Medical. Describes some special uses for parts of the radio spectrum that can be shared between multiple (often low-power) users without interference. Many short-range applications, such as wi-fi computer networks, use ISM frequencies.

Isolation transformer: A transformer that provides DC isolation between a current source and its load while passing an AC current or signal. Such network/appliance isolation is desirable in power and telecommunications applications to improve safety.

Isotropic radiator: A theoretical antenna that radiates evenly in all directions. Used as the basis for antenna gain measurements and comparisons (dBi).

ITU: International Telecommunications Union. United Nations agency responsible for telecommunications standard setting, harmonising regulations and radio frequency spectrum allocation agreements. Holds World Radiocommunications Conference. http://www.itu.int

ITU Day: World Telecommunications and Information Society Day. Observed on May 17 each year, it marks the anniversary of the founding of the International Telecommunications Union and the signing of the first International Telegraph Convention in 1865. Amateur radio clubs sometimes operate special event stations on this date.

ITU regions: Geographically-based administrative regions used by the International Telecommunications Union to manage the radio spectrum. Region 1 is Europe, Russia and Africa, Region 2 is the Americas while Region 3 is non-Russian Asia and Australasia.

ITU zones: A geographically-based zoning system that divides the world into 90 zones. These zones are used by amateurs as geographic markers for some awards and contests.

J

j: symbol often used to denote the reactive (or imaginary number) part of a complex impedance.

J-pole: J antenna. A type of vertical antenna commonly used on VHF and UHF bands. Comprises a 1/2 wavelength radiator next to a ¼ wavelength feedline tuning section.

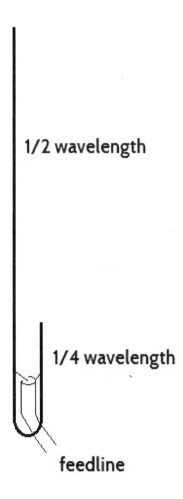

Jammer: A person or thing that causes deliberate interference to communication. May be a pirate or misbehaving licensed amateur.

JFET: Junction Field Effect Transistor. A special type of transistor often used in small signal audio and radio frequency circuitry, especially where a high input impedance is required. Unlike BJTs, JFETs are voltage rather than current operated.

JS8 Call: A version of FT8 that allows keyboard to keyboard conversations under weak signal conditions. Previously called *FT8 Call*.

JT9: A weak signal very narrow bandwidth digital mode developed by Joe Taylor K1JT. Intended for HF use.

JT65: A weak signal narrow bandwidth digital mode developed by Joe Taylor K1JT.

Junction transistor: The most common type of transistor. See *BJT*.

Junk: Old hoarded electronic equipment that takes up space. Owners typically have the best intentions of selling it at hamfests, giving it away or throwing it out. But they rarely do due to the fear that it will be needed the week after it's discarded.

Junk box: The box (which could be a shelf, the floor or a whole shed) in which items too good to throw out are hoarded in the hope that they will become useful for a future project.

K

K (diode): label for cathode connection on a diode.

K (Morse): Sent at the end of a Morse code transmission as an invitation for any station to transmit.

K-index: A short-term measure of geomagnetic activity that gives an indication of disturbance to the ionosphere. It is expressed in a range from 0 (quiet) to 9 (very major storm). Kp index is a global value based on readings from multiple observatories. The K index is often converted to an A index as a longer-term measure.

Kenwood: A popular Japanese manufacturer of amateur radio equipment. http://www.kenwood.com

Kerchunk: The act of keying up a repeater without identifying.

Key (Morse): A hand-operated up-down momentary switch connected to a transmitter or audio oscillator for sending Morse code. Character and interval spacing is manual so no electronic keyer timing circuit is required.

Key (to transmit): Key the transmitter. Refers to pressing the PTT on a microphone or otherwise activating a radio transmitter. Keying may cause internal circuitry to switch over, and on most modes (except SSB when nothing is being said) cause a signal to be radiated.

Key clicks: Spurious impulse signals generated above and below the transmitting frequency by a malfunctioning or poorly built CW transmitter.

Keyer: An electronic item, used in conjunction with a paddle, to provide accurate timing of dits and dahs when sending Morse code. May be a separate unit, or, more frequently now, built in to a transceiver.

Key up: To activate a repeater by transmitting a signal on its input frequency (with subtone if required).

Kilo: $x10^3$. Times one thousand. E.g. 1 kilovolt (kV) = 1,000 volts.

Kit: A package of parts with instructions that you assemble to make something that (hopefully) works and is useful. Can save money and be educational.

KN: Sent at the end of a Morse code transmission as an invitation for only the called station to transmit. The letters K and N are sent run together with no gap between them. (CW abbreviation)

Knife-edge diffraction: Refers to spreading of light or radio signals due to being passed across an edge such as a mountain range. This spreading can bend VHF/UHF radio signals so that they can be unexpectedly heard in valleys.

Knife switch: An open type of heavy-duty switch often used to change over antenna connections, especially those involving open wire feedline.

Koch method: An approach to learning Morse code that starts with learning two characters at the desired speed. When the learner gets proficient with these more characters are added.

L

L: Refers to inductance or inductors.

L-match: A form of antenna coupler popular with portable and low power operators. Suitable for end-fed wire antennas.

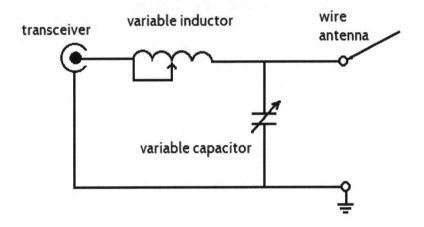

L-network: Inductor and capacitor circuit commonly used to transform impedances such as required between stages in a radio circuit or transceiver and antenna. Also *L-match*.

Ladder filter: A type of crystal filter using several crystals of the same frequency connected in series with capacitors to earth. Commonly used in SSB equipment to provide receiver selectivity and transmitter opposite sideband suppression.

Ladder line: A type of balanced antenna feedline used for some types of multiband HF dipoles. The spacers between the two wires make it look like a ladder. With a 300 to 450 ohm characteristic impedance, it is closely related to *open wire feedline*.

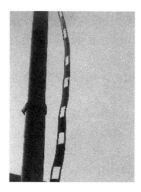

Latching relay: A special type of relay with extra contacts that allows its other contacts to remain in their energised position even after the voltage that triggered it has been removed from the coil.

Lattice filter: A type of crystal filter using several crystals of differing frequencies to provide the main selectivity determining element in FM and SSB receivers.

LC ratio: Inductance capacitance ratio. In a tuned circuit the inductance (in H) divided by capacitance (F). Or, more likely in radio circuits, uH/uF.

LCD: Liquid crystal display. Commonly used on transceiver frequency, signal strength and control status readouts.

LDR: Light dependent resistor. Component that varies resistance according to light exposure. Used for lights that switch on at night.

Lead acid: A type of wet cell rechargeable battery commonly used in cars. Sealed lead acid typed are sometimes used by portable amateur radio operators but are being supplanted by lighter but dearer lithium-based batteries.

Lead free: Solder without lead. Compulsory in Europe due to safety directives. Has a higher melting point than conventional lead/tin solder.

Leaded (components): Refers to electronic components that have leads, i.e. through-hole and not surface-mount.

Leaded (solder): Refers to solder containing lead.

Leakage: Refers to unwanted RF energy being emitted from a device such as a microwave oven or stage in a transmitter or receiver. Often due to poor shielding or construction practice, leakage can spoil the performance of equipment or pose a safety risk.

Leaky: Refers to a battery, capacitor or other component that is oozing fluid or conducting electricity when it shouldn't. This is a sign that it is defective and should be replaced.

Leaky grid: Type of detector common in early tube TRF or regenerative receiver circuits. Also grid-leak detector.

Lecher line: An early piece of radio test apparatus, comprising two parallel metal rods and a moveable shorting bar that allows one to measure the wavelength of signals at UHF and microwave frequencies with the help of an indicator meter or lamp whose brightness varied as the bar was slid. Because frequency is easily derived from wavelength, lecher lines were able to measure frequencies. Newer techniques for this are now available but lecher lines remain useful today to transform impedances, explain transmission lines and demonstrate the existence of standing waves.

LED: Light emitting diode. A diode that emits light when current is applied. Commonly used in front panel indicator lamps or torches.

LEO: Low earth orbiting. Refers to a satellite that has been launched to orbit close to earth. Signals from such satellites are strong but brief due to the speed of orbit relative to earth and consequent shortness of passes.

LF: Low frequency. The 30 to 300 kHz part of the spectrum. These frequencies are used for broadcasting in Europe and navigation purposes elsewhere. Known for its large antennas and long-distance ground-wave coverage. While commercial use of the band has declined, amateurs have recently gained some LF allocations.

Licence: Permission to operate on amateur frequencies granted by government after passing an examination in radio theory and regulation, and, in some cases, paying an annual fee. Most countries have two or three levels granting successively more privileges.

Lid: A poor or discourteous operator. (USA)

LiFePO4: Lithium Iron Phosphate. The chemical composition of a type of lightweight rechargeable battery. LiFePO4 batteries are popular amongst amateurs who operate portable. Less energy dense but safer than *lithium ion* batteries.

Lift: Propagation enhancement that boost signals and extends communication range. The term is most used to describe tropospheric ducting on VHF and UHF.

Lightning arrester: A component installed in the antenna lead that protects feedlines and equipment from high voltages such as may be present from a nearby thunderstorm. Lightning arresters contain an air gap between the antenna connection and a connection to earth. This has no effect under normal circumstances. But when high voltages are present, such as from a thunderstorm, the gap breaks down and provides an alternative path to ground for the energy picked up. Beneficial as they may be, lightning arresters are no substitute for disconnecting your equipment from the antenna when storms are forecast. Neither will they protect against a direct hit.

Line of sight: Refers to a signal path without physical obstruction between transmitting and receiving stations. Such paths should be easily spanned on VHF and UHF frequencies with low transmit powers. It was initially thought that VHF and UHF signals were only capable of line of sight communication but improved equipment and techniques soon allowed communication over more obstructed paths.

Linear amplifier: A type of transmitter RF power amplifier with an output power directly proportional to driving power as required for modes such as SSB. Commonly connected to 100-watt transceivers to allow operation near the legal power limit or low power equipment to deliver 30 to 100 watts.

Linear taper: Refers to resistance varying proportionately to a shaft's degrees of rotation such as occurs with some types of variable resistor. The other common type is *logarithmic taper*.

Linear translator (or transponder): A repeater like device that receives a band of signals on one frequency range and retransmits them on another frequency range without changing their characteristics (hence 'linear'). Able to handle all modes, linear translators are used in some communications satellites.

Link dipole: A half wave wire dipole antenna with switchable connections along the element to allow operation on multiple bands higher in frequency. Often used in inverted-vee configuration, lowering the centre allows access to the wire and the band to be changed. Suitable for the portable operator who does not need to change bands very often. The example below can be switched between 40, 30 and 20 metres, with 15 metres also possible if the entire antenna is used as a 3/2 wave dipole.

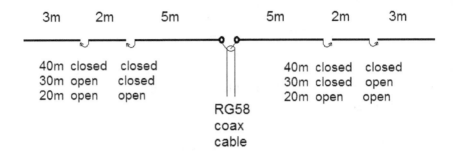

Linked repeater: A repeater that is linked to another via a radio, telephone or internet connection. Signals received by one repeater are transmitted by both repeaters, thus extending communications range. Links may either be permanent or switchable.

Linux: An open source operating system popular with computer geeks and programmers based on UNIX. As it requires less processing power than Windows it is often used on older slower computers that would otherwise be obsolete. http://www.linux.org

LiPo: Lithium polymer. A lightweight type of rechargeable battery often used to power portable equipment. Capable of delivering high currents from a small package, LiPo batteries need to be treated carefully to prevent mishaps.

List: A list of stations wishing to work a DX station, such as gathered by the controller of a DX net.

Lithium: An element used in lightweight high-capacity batteries.

Lithium ion: A lightweight type of rechargeable battery commonly used in mobile phones, handheld transceivers and other portable equipment.

LNA: Low noise amplifier. A high-performance amplifier that contributes very little noise to the signal it is amplifying. Often used at or near the feed point of microwave antennas used for satellite or SHF signal reception.

Load: Refers to something that is connected to the output of an amplifier or signal source. For example, a speaker for an audio amplifier or an antenna for a transmitter. The load should be impedance matched to the driving stage otherwise power transfer will be inefficient and equipment damage may result.

Load down: What can happen when a low impedance load is connected across a high impedance part of a circuit. For example, if 8-ohm headphones are connected directly across a crystal set's high impedance headphone socket. The headphones' low impedance will short nearly all of the signal to earth and nothing will be heard. The solution is to use high impedance headphones or install a high to low impedance step-down transformer to provide efficient power transfer.

Load up: Cause a piece of metal to act as an antenna. Unless you are very lucky this normally requires an antenna coupler for impedance matching.

Loading coil: Inductor used in an antenna to reduce its resonant frequency and allow operation on bands that you otherwise would not have room for. The loading coil can be located at different positions along the antenna - see *base loaded*, *centre loaded* and *top loaded*.

Lobe: A direction in which a directional antenna has a concentration of radiation. Opposite to *null*.

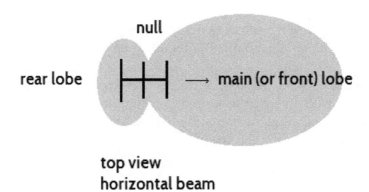

Local oscillator: A stage in a radio receiver that generates a low-level radio frequency signal. Required for processes such as mixing, converting and detecting. A receiver or transceiver may have several local oscillators, some of which go by different names e.g. *beat frequency oscillator* or *carrier oscillator*.

Lock: A button that when pressed disables important controls (such as transceiver tuning knobs) from operating. This feature is useful to protect against inadvertent frequency shifts when the transceiver or tuning knob is bumped.

Log (or logbook): A written record of an amateur station's on-air activity including contacts made. Previously compulsory logging is now voluntary in most countries but many amateurs still keep a log, especially if they are chasing awards. Logs were previously only done on paper but are now also computerised with a wide variety of programs available. Many active amateurs have their log online so you can quickly verify that you made contact with them.

Logarithmic taper: Refers to a potentiometer whose resistance changes disproportionately to the degree rotation of its shaft according to a logarithmic curve. This characteristic is typical of variable resistors used for volume or level controls as it allows easier adjustment of volume at lower levels than a linear taper would.

Log periodic: A type of beam antenna that exhibits a much wider bandwidth than a yagi. Instead of covering only a single amateur band, a log periodic may provide directional gain over a 2:1 frequency range, allowing multiple bands to be covered. Trade-offs include higher material cost, greater weight and lower gain than a monoband yagi. Log periodic beams are most used for VHF/UHF TV reception and commercial or defence HF communication.

Logic gate: A digital circuit that produces a 0 or 1 output dependent on the levels present at its (usually) two inputs. The outputs from all combinations of inputs are contained in a truth table. Logic gates are available for different functions including and, nand, or, nor, exclusive or & exclusive nor. ICs containing four of the same type of gate are commonly available. Correct operation can be tested with a *voltmeter* or *logic probe*.

Logic probe: A piece of electronic test equipment used to determine the logic state (i.e. 0 or 1) of digital circuitry. This is very useful when constructing or troubleshooting.

Long path: Refers to HF signals going the long way around the globe between two stations. This is most likely where stations are close to being antipodes, for instance Europe and Australia.

Despite the longer distance the long path may offer strong signals due to a favourable combination of the sun's position relative to the communications path, frequency and propagation. Opposite to *short path*.

Long wave: An alternative term for low frequency, LF or 30 – 300 kHz.

Long wire: A type of antenna, typically end-fed, used on the MF and HF bands. Strictly speaking it is longer than 1 wavelength at the operating frequency. The length is often not specified and an antenna coupler will likely be needed at the transmitter end.

Loop: A type of antenna. There are two main types – large and small. An example of a large loop is a circle, square or triangle of wire approximately 1 wavelength perimeter. A small loop, often called a 'magnetic loop', comprises a much smaller loop of thick copper or aluminium brought to resonance with a variable capacitor. Another small type of loop, designed mainly for LF and MF receiving, have several wire turns on a frame. Loops are well-known for their sharp nulls and low noise pick-up so are often used for receiving and direction-finding.

Loopstick: A coil of wire wound on a ferrite rod such as used as the antenna in AM broadcast band receivers.

LOS: Loss of signal. Refers to when an orbiting satellite drops below the horizon and its signal vanishes. Opposite to *AOS*.

Loss: The reduction in a signal, voltage or power level due to inefficiency in a circuit, antenna, power lead or transmission line. Often dissipated as heat.

LOTW: Logbook of the world. A system that permits electronic confirmation of contact such as may be required for awards. Sponsored by the *ARRL*. http://www.arrl.org/logbook-of-the-world

Low angle radiator: Refers to an HF transmitting antenna that sends most of its energy at low angles towards the horizon parallel to the earth's surface. This is useful for long distance HF communication. Examples include high horizontal dipoles and verticals erected over a conductive ground. Opposite to *high angle radiator*.

Low-Z: Low impedance. Refers to low impedance circuits, antennas or headphones. In other words those where signal voltages are relatively low but currents are relatively high.

Lowfer: Low power, low frequency radio especially the 160 – 190 kHz licence-free segment available in the USA. (USA)

LPF: Low pass filter. A filter designed to pass signals of less than a certain frequency but reject signals above that frequency. Using capacitors and inductors in a pi-network configuration such filters attenuate any higher frequency harmonics and lessen the risk of interference to other spectrum users. The most common LPF filter type has a 30 MHz cut-off to lessen VHF TV and radio interference caused by harmonics from nearby HF equipment. Low pass filters can also be used in the audio stages of SSB receivers to improve reception of CW signals (which have a narrower bandwidth than voice signals).

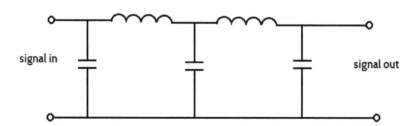

signal in

signal out

LSB: Lower sideband. An SSB transmission where the audio component extends below the frequency of the suppressed carrier. Normally used on amateur bands below 10 MHz. Opposite to *USB*.

LT: Low tension. Low voltage such as required for filament connections in tube equipment.

LUF: Lowest usable frequency. The frequency below which satisfactory HF communication via the ionosphere will be lost due to increasing D-layer absorption. Not a hard and fast value but depends on factors such as transmitter output power, antenna gain, receiving site noise and modulation efficiency. Compare with *MUF*.

Lug: A terminal designed to allow a cable to connect to a metal case, chassis or circuit board. A lug normally has a solder hole or crimp connection for the wire at one end.

M

m: metre. Used to describe amateur bands and other frequencies in terms of wavelength. E.g. 20, 17 and 15 metre amateur bands. Or 49, 41 and 31 metre short wave broadcast bands. 300 divided by metres will give the approximate frequency in megahertz.

Magnetic loop: A compact type of HF antenna characterised by small size and narrow bandwidth. Magnetic loops typically use thick copper or aluminium with a circumference of 0.1 to 0.2 wavelength on the operating frequencies. A high voltage variable capacitor allows the loop to be tuned to the exact operating frequency. Used by hams with little space for conventional dipoles, verticals and loops.

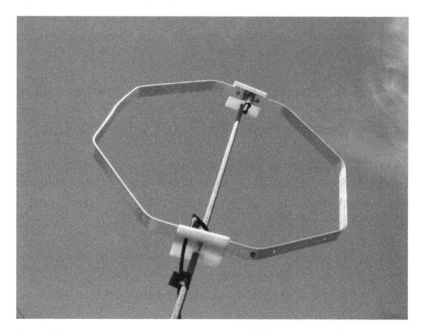

Magnet mount: A magnet on the base on a VHF or UHF mobile vertical antenna that allows easy attachment to a vehicle roof.

Maidenhead locator: Another name for *grid locator*, named after the English town in which the system was adopted in 1980.

Male: Refers to a plug or pin that plugs in to a (female) socket or receptacle to form an electrical connection. While it is most common for female connectors to be on equipment and males to be on cables, this is not always the case. For example, many older CB and ham HF transceivers had a 4 or 8 pin male socket (i.e. recessed pins) into which a female at the end of the microphone's curly cord plugged in to.

Marconi antenna: A class of antenna typically comprising a vertical element of resonant length mounted over a conductive ground plane or radials. Named after its inventor.

Maritime mobile: Operating amateur radio from a vessel at sea.

MARS: Military Auxiliary Radio Service. A US Department of Defense program comprising civilian amateur radio operators who provide communications support for military forces. https://www.mars.af.mil (USA)

Mast: A pole for supporting antennas. A less emotive term than 'tower' the term is preferred when negotiating permission to erect with spouses, homeowner associations or municipal councils.

Masthead amplifier: A radio frequency amplifier installed at the antenna, before the feedline, to improve receiver sensitivity and overcome feedline loss. Used for long distance VHF and UHF radio and TV reception.

Matching network: One or more capacitors and/or inductors connected to transform impedances and allow signals to efficiently pass between circuit stages or from a transceiver to an antenna.

Matrix board: A type of circuit board, used mainly for hobby and prototype projects, that has holes punched in a grid. Hole spacing is typically 2.5mm to easily accept component and IC leads.

Mayday: A distress call made by radio. Other communication must cease and stand by to assist if required.

MDS: Minimum detectable signal. The weakest signal that a receiver can detect to produce a signal of a given strength.

Mercury switch: An enclosed glass switch with a blob of mercury inside. Tilting the glass allows the mercury to bridge the switch contacts, closing the circuit.

MF: Medium frequency. The 300 kHz to 3 MHz part of the spectrum. Used for AM broadcasting, especially in North America and Australia. Amateurs have MF allocations at 472 kHz and 1.8 MHz. Provides local ground wave coverage with extended distances at night due to ionospheric propagation.

Mechanical filter: A selective filter, often passing frequencies around 455 kHz, sometimes used in SSB transmitters and receivers to filter out the unwanted opposite sideband. Other types of mechanical filters have even narrower bandwidths such as desired for clear reception of Morse code signals.

Medium wave: See *MF*.

Mega: $\times 10^6$. Times one million. E.g. 1 megahertz (MHz) = 1,000,000 Hz.

Memory: Memory channel. A feature in a transceiver or receiver where you can store commonly used frequencies and other settings so you can quickly return to them later. Memory channels can often be scanned, with the receiver stopping on busy frequencies and scanning resuming when activity has ceased.

Memory effect: An effect some claim exists with nickel cadmium and nickel metal hydride rechargeable batteries where, if you do not fully discharge the battery, it will have diminished capacity after subsequent chargings.

Menu: A way of allowing lesser used functions or settings on a transceiver to be adjusted without requiring a dedicated knob or button for each one. These are accessed by pressing a menu button that reveals sets of different functions for buttons or controls and scrolling through them with another control. A common source of frustration amongst larger fingered amateurs, menus are most common on small equipment such as handheld and mobile transceivers.

Metal film: A high quality type of resistor with a thin metal layer to form the resistance. Superior to the cheaper carbon film type, metal film resistors are desirable in low noise low signal stages of circuits such as audio amplifiers and test instruments.

Meteor scatter: A form of VHF or UHF long-distance communication that relies on bouncing signals from meteors.

Meter movement: A mechanical device, comprising a magnet and a moving coil of wire connected to a pointer, that indicates the current flowing through a circuit. Commonly used on radio equipment to indicate received signal strength or transmitter output power.

MFJ: US manufacturer and distributor of antennas and accessories for the amateur. http://mfjenterprises.com

Mica capacitor: An old-fashioned non-polarised capacitor used in applications where low values of capacitance are required such as in radio circuitry. High voltage ratings make them suitable for RF power amplifiers. A variant is silver mica which is good for critical, frequency-sensitive applications. Typical values are 10 pF to 1 nF.

Mica washer: A clear electrically insulating washer typically mounted between a power transistor and its heatsink. This allows thermal conductivity but electrical insulation between the component and its heat sink.

Microcontroller: A basic computer, often on a small circuit board, that can be programmed to read inputs, generate signals and control electronic circuits.

Microprocessor: An IC that contains all the parts of a computer's central processing unit.

Microphone: The transducer you speak in to when talking on a transmitter.

Microphone amplifier: A stage in a transmitter than amplifies the sound input from the microphone to a level sufficient for subsequent stages.

Microphone gain: A setting that determines the gain of the microphone amplifier. Too low a setting results in you not being heard while too high a setting causes transmission of extraneous noise.

Micro: $x10^{-6}$. One millionth. E.g. microfarad (uF) or microhenry (uH).

Microsat: A small, often low earth orbiting, artificial satellite typically weighing under 100kg.

Microswitch: A small momentary switch often controlled by a lever such as common in electromechanical devices such as CD players and video tape recorders.

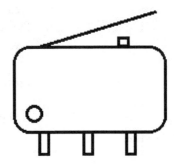

Microwaves: Normally refers to frequencies above about 1 GHz (1000 MHz).

Milli: $\times 10^{-3}$. One thousandth. E.g. milliamp (mA) or millihenry (mH).

MININEC: Popular software for antenna modelling.

Mixer: One of the basic stages in radio circuits. A mixer has two inputs and one output. Different signals are applied to each input. At the output are two frequencies. One is the sum of the frequencies of the inputs while the other is the difference of the frequencies of the inputs. Only one is normally required so a filter is added to pass only the desired frequency. Mixers have numerous uses including balanced modulators in SSB transmitters and product detectors in superhet or direct conversion receivers.

MM: Maritime mobile. Operating amateur radio from on board a vessel.

MMANA-GAL: Popular software for antenna modelling. Free for hams. http://hamsoft.ca/pages/mmana-gal.php

MMSSTV: A common computer program for transmitting and receiving slow scan television. http://www.hamsoft.ca/pages/mmsstv.php

Mobile: Refers to an amateur radio station that is being operated while on the move. Most likely in a car but could also be in a train, bus, boat or while walking along. Some countries require that amateurs operating away from home identify as mobile.

Mobile antenna: An antenna designed to be mounted on a vehicle. HF mobile antennas are typically helically-wound verticals while VHF/UHF mobile antennas are normally a quarter or five eighth wavelength vertical.

Mobile flutter: Fast fading of a signal due to the path moving caused by the transmitting and/or receiving station moving. Also *flutter*.

Mobile mounting bracket: A normally U-shaped bracket that allows a transceiver to be mounted inside a vehicle.

Mod: A modification to equipment to improve performance or add a wanted feature.

Mode: A method of amateur radio communication. Examples include SSB, AM and FM (analogue voice modes), DSTAR (a digital voice mode), FT8, WSPR and PSK31 (digital modes), CW (Morse code) and SSTV (a means of transmitting pictures). Some modes require a specialist transceiver for that mode while others simply require a computer program to be downloaded and a connection be made between the computer and the transceiver.

Modem: Short for modulator / demodulator. Connected between a computer terminal and a radio transceiver (or telephone line) to enable data communications.

Modulation: The process of imposing voice or other intelligence on a radio signal as done in a radio transmitter. Modulation methods include AM, FM, SSB and various digital modes.

Modulation monitor: A piece of test equipment used to monitor the quality of modulation from a speech transmitter.

Momentary switch (normally closed): A type of switch where contact is interrupted while pressing on it but is restored once released.

Momentary switch (normally open): A type of switch where contact is made while pressing on it but ends once released. An example is the push to talk button on a microphone or a Morse key.

Monoband: Refers to equipment or antennas capable of operation on one band only.

Monolithic capacitor: A non-polarised capacitor used for general purpose applications in electronic equipment. Typical values are 1nF to 100 nF.

Moon bounce: Long distance radio communication achieved by bouncing VHF or UHF signals off the moon. Also known as *EME*.

Morse code: The series of dits and dahs used to send messages by telegraphy. Now used to refer to the International Morse code, though this was not the system originally developed by Samuel FB Morse. Also see *CW*.

Morse requirement: A previous international convention that required amateurs to pass a test in Morse code before being allowed to use bands below 30 MHz.

MOSFET: Metal oxide silicon field effect transistor. A type of field effect transistor commonly used for high power switching and amplification including final amplifier stages in radio transmitters.

Motorboating: Refers to instability or oscillation that causes an audio amplifier or transmitter to make a pulsing noise. May be caused by feedback or a poor power supply. Switch off immediately!

MOV: Metal oxide varistor. A two-legged component that exhibits high resistance at low voltages but low resistance at high voltages.

Used in power supply filtering and over-voltage protection applications.

Moving coil microphone: See dynamic microphone.

Moxon: A type of compact beam antenna similar to a VK2ABQ. Comprises two wire elements with ends bent in on themselves to form an antenna that is narrower (if horizontal) or shorter (if vertical) than a conventional two element yagi. Named after the late Les Moxon G6XN.

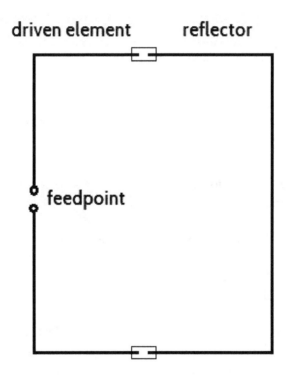

Mud duck: CB radio slang for a weak, difficult to copy signal buried under noise.

MUF: Maximum usable frequency. The maximum frequency that will support ionospheric propagation on a particular path. Dependent on factors such as solar conditions, time of day and distance between stations. See also *LUF* and *OWF*.

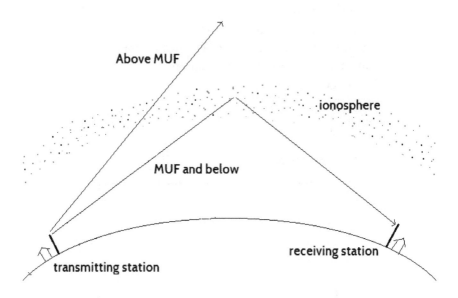

Multiband: Refers to an antenna or transceiver that operates on multiple bands.

Multi-hop: Relates to ionospheric propagation where, due to distance between the transmitter and receiver, the signals needs to return to earth one or more times before being once again reflected by the ionosphere.

Multimeter: A useful piece of electronic test equipment that measures voltage, current and resistance. May also measure capacitance, inductance and frequency. All hams should own one.

Multi-operator: Refers to a club, contest or DXpedition station where several people make contacts, often simultaneously on different bands. Opposite to single operator.

Multi-path: Relates to propagation where the signal may reach the receiving station from the transmitting station via multiple paths. Where the paths are of slightly different lengths this can affect the time signals arrive and cause echo effects.

Mute: To quieten. A control or button on radio receiver (especially AM or FM) that quietens the receiver unless a signal appears (known as "breaking the mute"). Also *squelch*.

Mx: Now rare abbreviation for metre with respect to amateur bands. E.g. 80mx, 40mx, 20mx etc. (archaic)

N

N-type: A style of plug and socket used for transceiver-antenna connections. Recommended for VHF and UHF.

Nano: $x10^{-9}$. One thousand-millionth. E.g. nanofarad (nF).

Nanosat: A small, often low earth orbiting, artificial satellite weighing between 1 and 10kg.

Narrow filter: A crystal, mechanical or audio filter used to improve reception of narrowband modes such as *CW*.

Narrowband: Of or occupying a small frequency bandwidth. For example, CW or slow speed digital modes. Or a magnetic loop antenna that works efficiently over only a few kilohertz unless re-adjusted. Opposite to wideband (or broadband).

National society: Describes the organisation in each country that represents amateur radio interests to government and helps fund international representation through the IARU. Often provides services such as a magazine, QSL Bureau, contests and club support. Many local clubs are affiliated with the national society.

NB: Noise blanker. Circuitry or software that suppresses noise while still allowing reception of the desired signal.

NBFM: Narrow band frequency modulation. Refers to FM with low deviation such as used for two-way radio communications. Opposite to *WBFM*, as used for high quality broadcasting.

NBTV: Narrow band television. Low definition television whose narrow bandwidth is similar to that of a voice transmission.

NBVM: Narrow band voice modulation: A communications mode allowing speech communications in a narrower than usual bandwidth.

NC: Normally closed. Relates to contacts on a switch or relay that are closed when in its un-pressed or un-energised state.

NDB: Non-directional beacon. Beacons, typically transmitting in the 200 – 500 kHz range, used for navigation. Many have closed down as other systems took over.

Negative earth: Refers to a wiring system where the negative connection of a battery or power supply is connected to chassis, ground, a metal case or a vehicle's housing. This arrangement is convenient for auto electrics and devices using NPN transistors (which are most common). Opposite to *positive earth*.

Negative feedback: Refers to an amplifier stage that has had a portion of its output fed back to its input to decrease its gain. This is sometimes deliberately done to reduce distortion.

Negative resistance: A phenomenon where a circuit or component appears to defy Ohms law – that is an increase in voltage across it results in a decrease in electrical current through it. Examples of components exhibiting negative resistance include certain types of diodes and neon lamps.

Neon lamp: A small low-current lamp containing neon gas that requires about 90 volts to operate. Commonly used as a power indicator for mains-powered equipment it can also be used to detect RF at high impedance points along antennas.

Net: An on-air gathering on a regular frequency at an appointed time. May be daily or weekly. Often run by radio clubs or special interest groups.

Net controller: The station that runs a net. Duties include calling for check ins, making a list of stations who have called in, and keeping order so that everyone gets a fair say and there is no *doubling*.

Network analyser: A piece of RF test equipment that allows measurement of various electrical characteristics of components and assemblies such as amplifiers and band pass filters. Comprising a signal source and multiple receivers, tests are done by analysing the component's response when a signal is applied.

Neutralisation: An internal adjustment often required for high-gain vacuum tube RF power amplifier stages to prevent instability.

NF: Noise figure. A number used to indicate the noise performance of an amplifier or sensitivity of a receiver, especially on UHF and microwave frequencies. Expressed in decibels, the lower the figure the less internal noise the device or stage contributes.

Nibbling tool: A useful hand-held tool that allows small chunks to be punched from metal chassis or printed circuit board material. This allows square cut-outs in panels to be made for switches, LCD displays, meter movements and large sockets.

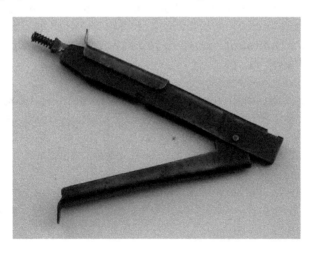

NiCd: Nickel cadmium. Older style of rechargeable battery that has largely been replaced with NiMH and lithium-based types. However their low internal resistance still makes them useful for high current applications.

Nil: Sent to indicate that not a trace of your transmission has been received, as may occur if band conditions have changed. (CW abbreviation)

NiMH: Nickel metal hydride. A style of rechargeable battery that largely replaced NiCd batteries and is in turn gradually being superseded by lithium-based batteries. Regular types are prone to self-discharge when not in use but low discharge varieties are available.

NO: Normally open. Relates to contacts on a switch or relay that are open circuit unless the switch or relay is actuated. Opposite to normally closed (*NC*).

No code: Used to refer to a category of amateur licence that did not require a Morse test. Before the Morse requirement was dropped in the early 2000s, no-code licenses in most countries confined their holders to frequencies above 30 MHz.

Node: A connection point in a data communications network that can originate, receive, store or send information along defined paths.

Noise: Any unwanted signal that detracts from the purity of received or amplified signals. May be generated internally within a microphone, receiver or amplifier or picked up from outside (e.g. band noise or lightning static).

Noise blanker: Circuitry in a receiver designed to suppress unwanted RF noise such as radiated from power supplies, power lines and motor vehicles. Vary greatly in effectiveness.

Noise bridge: A piece of test equipment often used to measure complex antenna impedances in conjunction with a broadband RF noise generator and calibrated receiver. Mostly now replaced by antenna analysers.

Noise cancelling microphone: A microphone arranged to emphasise speech and cancel extraneous noise from other directions. Often used in mobile applications.

Noise floor: Relates to the level of ambient RF noise at a location. A high noise floor, such as in a large city, means that you are unable to hear weak signals. A low noise floor, such as in the wilderness, makes weak signals intelligible.

Non-ionising radiation: Electromagnetic radiation that causes heating but is unable to change the atomic structure of the things it heats. RF energy is an example. Many countries have exposure standards for non-ionising radiation.

Non-linear: Refers to an RF amplifier stage (e.g. Class C) whose output power is not directly in proportion to the strength of the signal applied. This is tolerable for some modes (such as CW or FM) but not others (e.g. SSB). Non-linear amplifiers are used due to their higher efficiency and slightly simpler construction.

Non-resonant antenna: An antenna that presents a complex impedance, containing either capacitive or inductive reactive components, on the desired operating frequency. It can only be used if a matching network is added to tune out reactance and make any required impedance transformation. Such a network can either be placed at the antenna feed point, or, if a low-loss feedline is used, take the form of an antenna coupler at the transceiver end. Opposite to *resonant antenna*.

Notch filter: A type of filter that is designed to reject (or 'notch') signals of a particular frequency. This can be useful if trying to separate signals from interference.

Note: Refers to the quality of sound from a Morse code transmitter. Normally expressed in terms of the tone report in the *RST* scale.

Novice licence: The name formerly used for the entry level amateur licence in many countries. Typically provided low power access to a limited selection of bands. Replaced by Foundation or Technician licence in many countries.

NPN: Relates to a bipolar junction transistor whose collector is more positive than the emitter. This is the most common type of transistor used, possibly due to the prevalence of negative earth circuits and systems. Opposite to *PNP*.

NR: Noise reduction. Circuitry in a receiver to assist reception by reducing unwanted noise. A variety of techniques are employed, ranging from simple diode noise clipping to advanced digital signal processing.

NTS: National Traffic System. An organised system of passing messages via amateur radio sponsored by the ARRL. http://www.arrl.org/nts (USA)

Null: The low or zero point in an adjustment, indication or antenna radiation pattern. Often reached when a bridge circuit is adjusted to balance. A null may be indicated by a meter reading dropping to zero, a light going out, a sound disappearing or a low point in signal when a directional antenna is turned.

NVIS: Near Vertical Incidence Skywave. A term used to refer to short and medium distance HF communication using the ionosphere for reflection. This can be useful for contacts beyond HF groundwave or VHF/UHF range. Best exploited with an antenna with significant high-angle radiation such as a low dipole. Frequency selection depends on time of day and phase of the solar cycle. Frequencies that are too low will be absorbed while those too high will skip over the stations you wish to contact. 3.5, 5 and (sometimes) 7 MHz are useful for this work.

O

OCF dipole: Off-centre fed dipole. A half wavelength of wire fed part way along it (instead of at the centre as with the standard centre-fed half wave dipole). Off-centre feeding has a number of features attractive to amateurs including operation on popular harmonically related HF bands (3.5, 7, 14 MHz etc). Feeding away from the centre raises impedance which means that a step-up transformer is required if feeding with 50-ohm coaxial cable.

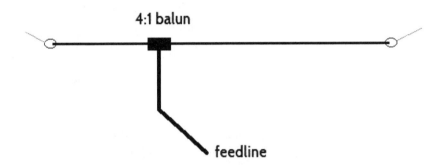

Octal: An 8-pin style of tube socket used in the 1940s and 1950s. The tubes typically had a Bakelite base attached to the glass envelope. Later styles, mostly with 7 or 9 pins, were all-glass construction.

Off-frequency: Refers to a transmitter or receiver not being on an agreed frequency. Depending on the transmitting mode this can result in distorted reception or no signal at all.

Offset: Repeater offset. Split. Refers to the difference between a repeater's receive and transmit frequency. Offsets are normally standardised in a country or region, for instance 600 kHz for many two metre repeaters.

Off-grid: Refers to being able to operate your station independently of the mains power grid by having your own solar, wind or other power source.

Ohm: The unit of electrical resistance represented by the Greek letter omega (Ω). Also kiloohm (a thousand ohms) and megohm (a million ohms).

Ohms Law: Defines the mathematical relationship between voltage, current and resistance in a circuit.

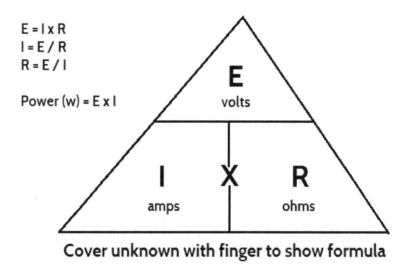

$E = I \times R$
$I = E / R$
$R = E / I$

Power (w) = E x I

Cover unknown with finger to show formula

Olivia: a narrow bandwidth weak signal digital mode that allows keyboard to keyboard chats. http://www.oliviamode.com/

OM: Old Man. Any male radio amateur (regardless of age).

Omnidirectional: Describes an antenna that has equal radiation in all directions, such as the theoretical isotropic radiator. In common use it may also refer to antennas that have an equal radiation in all azimuth directions, such as a ground plane or vertical dipole, even though they may have nulls at certain elevations.

On the side: A CB radio-derived expression for a station who was previously in an on-air conversation but has now signed and is now just listening.

Online receiver: A receiver able to be controlled via the internet. Handy for monitoring your own transmission or checking reception conditions in other areas. Also *web receiver*.

OO: Official Observer. ARRL-appointed volunteer who monitors bands and informally advises amateurs of dirty signals or poor operating habits as a means of amateur self-regulation. (USA)

Op: Operator. Person who operates an amateur radio station. (CW abbreviation)

Op amp: Operational amplifier. A voltage amplifier circuit, normally contained in an IC, that allows its gain and other characteristics to be set by changing external component values. Features two inputs (inverting and non-inverting) and one output. Often used in low level audio amplifiers, audio filters and test equipment.

Open (circuit): Refers to two contacts or connections through which current cannot flow as there is no electrical path between them.

Open (frequency): Refers to a frequency or band that supports radio communication between two particular points. Openings depend on factors such as the season, time of day and solar conditions.

Open repeater: A repeater whose sponsors welcome usage by all radio amateurs in an area. Opposite to closed repeater. (USA)

Open wire feedline: A form of balanced feedline comprising two wires separated by spacers. Typically has a high impedance (e.g. 300 – 800 ohm).

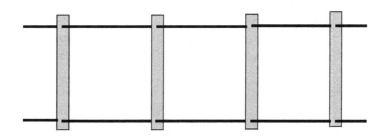

Opening: An occurrence of radio signal propagation, especially to a distant or rarely contacted location for the frequency band used.

Operate: To use an amateur radio station.

Operating privileges: Refers to what one is allowed to do on the amateur bands. Includes frequencies, modes and maximum transmitter output power limit. Varies by country and licence level.

Operator: Licensed operator. Someone qualified to use an amateur radio station.

Optical communications: Communication using light to convey information.

Optocoupler: Electronic component comprising a light emitting diode placed near a light sensor in a lightproof package. Sometimes used for some switching applications in preference to relays, optocouplers can provide perfect electrical isolation between two parts of a circuit.

OSCAR: Orbiting Satellite Carrying Amateur Radio. Successfully launched satellites are given an OSCAR number to allow easy identification.

Oscillator: An important radio circuit building block that generates an audio or radio frequency signal. Can be free-running (good for agility but bad for stability), crystal control (good for stability but bad for agility) or synthesised (good for both but more complex). Turning an oscillator on and off allows Morse code to be sent. Adding RF amplifier stages increases its output to form a real transmitter able to be heard at distance. Oscillators are also used in receivers – see *local oscillator*.

Oscilloscope: Piece of electronic test equipment with a screen to monitor electrical signal waveforms for level and distortion.

OT: Old Timer. A person who has been in radio for many years.

OTHR: Over the horizon radar. Radar that uses HF frequencies to allow detection of objects beyond range of conventional UHF radar systems. Can interfere with HF communications including amateurs.

Out: Sometimes said at the end of a station's last transmission in a conversation. Unlike 'Over', the speaker is not expecting a response. Therefore saying "over and out" is incorrect usage.

Out of phase: Refers to signals that are identical in waveform but are staggered in their timing. Two of them together can create odd effects such as signal echoes or cancellations. Splitting a signal into two components and making them out of phase with one another is sometimes used to make antennas directive or generate single sideband transmissions. Also see *phase difference.*

Output frequency: The frequency that a repeater transmits on. To use a repeater you must listen on the output frequency and transmit on its input frequency.

Over: Often said at the end of a voice transmission to signify that it is another person's turn to transmit. Also refers to a person's voice transmission (e.g. 'long overs' are long transmissions).

Over-deviating: Refers to an FM voice transmission whose excursion from its centre frequency is excessive for the receiver it is used with. This causes the signal to drop out on voice peaks. Talking quieter into the microphone is a short-term remedy but the real cure involves an adjustment or modification to the transmitter.

Over-driving: Relates to a condition where excessive audio or radio frequency energy is being fed to a stage in a transmitter or receiver, causing its output to distort. This can make signals less readable and cause interference to others.

Over-modulating: Relates to a condition where excessive applied audio is causing a transmitter to distort and potentially splatter.

Overtone mode: Relates to a crystal or crystal oscillator operating on a frequency that is an approximate third, fifth or seventh multiple of its fundamental frequency. This can enable a higher frequency signal to be obtained from a lower frequency crystal without frequency multipliers being required.

OWF: Optimum working frequency. The frequency that is considered optimum for reliable HF communication over a particular path. Dependent on factors such as solar conditions, time of day and distance between stations. Normally about 80% of the *MUF*.

P

PA: Power amplifier. The last RF amplifier stage in a transmitter before the antenna. Also *final amplifier*.

Packet radio: A form of data communication popular in the 1980s and 1990s. Predating widespread use of the internet it allowed live keyboard chats, personal messages and access to bulletin board systems via amateur radio. Data was provided in 'packets' that were automatically re-sent if not received correctly.

PACTOR: A 1990s digital mode that was at one time popular on HF. Using techniques from both packet radio and AMTOR, PACTOR is a robust means of exchanging data over a narrow bandwidth radio link.

Paddle: A hand-operated side-to-side switch for sending Morse code. Moving it one direction sends dots while dashes are sent in the opposite direction. Come in single and dual lever varieties. Most transceivers have circuitry that allows a paddle to be directly connected without external keyer control and timing circuitry. Paddles are favoured by many regular CW operators as they are less stressful to use than traditional up and down keys. See *iambic keyer*.

Paper entry: Refers to a contest log that has been written on paper and mailed (as opposed to an electronic log).

Parabolic: Dish-shaped. More precisely, having the shape of a parabola as defined by mathematical function. This shape is widely used for microwave radio antennas.

Parallel: Connecting components across one another. Opposite to *series*.

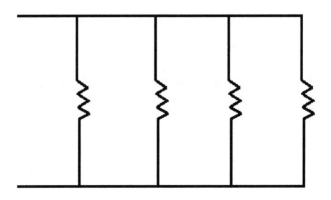

Parallel resonant: Relates to a tuned circuit with the capacitor and inductor connected in parallel. In this arrangement the impedance (i.e. resistance to radio frequencies) rises to a maximum at the resonant frequency with a fall off at frequencies either side. This characteristic has many uses in radio circuits. For example, parallel tuned circuits in a radio receiver or transmitter allow signals on the desired frequency to pass while acting as a virtual short circuit for signals on unwanted frequencies above and below. Another application is in trap dipole or vertical antennas where a parallel tuned circuit (or trap) part way along the antenna wire allows operation on multiple bands by automatically electrically lengthening or shortening the antenna, depending on the frequency applied. Also see *series resonant*.

Parasitic element: An element in an antenna that is not connected to the feedline such as a reflector or director in a beam antenna. Parasitic elements are added to provide directivity and gain over a single driven element.

Parasitic oscillation: An unintended oscillation in an amplifier stage caused by poor design, construction or adjustment. Parasitic oscillations reduce an amplifier's efficiency and cause spurious emissions.

Parrot repeater: A form of repeater that records a voice and retransmits it on the same frequency a few seconds later. This form of repeater needs only a single frequency and does not require separate transmit and receive antennas. However, they are only good for short transmissions.

Part 15: Refers to small radio transmitters that due to their low power do not require individual licenses. Term comes from the FCC's radio regulations. (USA)

Pass: The period of time that a satellite is within line of sight (and radio contact) of a particular location. Pass times for amateur satellites can be calculated with computer software or are made available via the *AMSAT* website.

Pass transistor: A power transistor used in the high current section of a regulated linear power supply such as required to run transmitting equipment. They are controlled by a voltage regulator that can only handle a small current.

Pass band: Relates to the range of frequencies admitted by an audio or radio frequency filter without attenuation. A wide setting allows wide bandwidth, high fidelity reception or reproduction while a narrow setting may reduce interference or hiss.

Passive circuit: A circuit that operates without requiring a power source or active components. Examples in RF design include filters, mixers and phase shift networks. Passive stages typically exhibit loss so usually need to be preceded or followed by an amplifier stage if appreciable output from them is expected.

Passive component: A component such as a resistor, capacitor or inductor that on its own cannot be used to generate or amplify a signal. Opposite to *active component*.

Passive repeater: A technique for improving reception of VHF or UHF signals over a non-line-of-sight path by receiving a desired signal on a hilltop with a high gain beam and retransmitting it on a second beam pointing into the valley. Alternatively, use may be made of a large reflective panel.

Path: The course a signal takes between stations in contact. This is usually the shortest distance between stations. Cases where it is not include contacts made by bouncing a VHF or UHF signal off a mountain's side or HF contacts where signals go the 'long path' around the world.

Path loss: The reduction in strength of a radio signal as it travels from origin to destination. Factors that affect path loss include distance, obstructions, antenna height, operating frequency and propagation conditions.

PBT: Passband tuning. An adjustment in advanced receivers that allow the receiver's pass band to be varied slightly in frequency to dodge interference.

PCB: PC board. Printed circuit board. Popular mounting arrangement for electronic parts where component leads are soldered to copper tracks that have been left behind in an etching process. PC boards comprise a sheet of fibreglass (or similar) covered by copper on one or both sides. Holes may be drilled for through-hole components. More complex printed circuit boards have multiple layers to allow closer spacing of parts.

Pedestrian mobile: Operating amateur radio when walking. Easy with a VHF/UHF handheld transceiver but harder on HF with bigger antennas required for usable efficiency. Photo below shows a 7 – 28 MHz magnetic loop.

PEP: Peak envelope power. Refers to the maximum power output reached by the signal from a transmitter as measured over one or more cycles when fully modulated. PEP is the same as average power for most modes except AM and SSB voice. An AM transmitter's PEP is typically rated four times its carrier power.

And on SSB PEP significantly exceeds average power, with the average power dependent on whether speech processing is used or not. PEP requires a special wattmeter, with an ability to capture and display short-term peak values reached, to accurately measure.

Phantom power: DC power as required by some types of microphone often delivered via the microphone cable. A similar concept is sometimes used to power masthead amplifiers by sending DC up the antenna feedline.

Phase difference: Refers to differences in the timing of signals, normally of identical frequencies. Phase difference is measured in degrees, with 45 degrees representing an eighth of a cycle, 90 degrees a quarter cycle and 180 degrees a half cycle out of phase. A 0-degree phase difference between two signals means that they are in step (or in phase) with one another, with no time difference. Awareness of phase differences and how they can cause signals to add or cancel one another out is important in some aspects of circuit design (e.g. phasing system SSB equipment), interference suppression equipment and certain antennas (e.g. phased dipole or vertical arrays).

Phase distortion: Refers to distortion commonly heard on AM shortwave signals due to variations in the path between the transmitter and receiver. Phase distortion can make signals unreadable if severe. Amateurs are fortunate in having SSB as a more robust mode less susceptible to its effects.

Phase modulation: A transmitting mode generated by imposing intelligence on a radio signal by varying the transmitter waveform's phase. Related to frequency modulation, a phase modulated signal can be heard on an FM receiver.

Phase noise: A form of noise caused by rapid fluctuations in a signal's phase. It may appear as distortion on the signal output of, for example, a local oscillator. This may affect the purity of signals heard on a receiver or sent by a transmitter. Phase noise is rarely noticeable as a problem unless trying to receive weak signals or seeking high quality audio reproduction.

Phase shift network: Circuitry that shifts the phase of an audio or RF signal by a certain amount such as required for phasing SSB generation or in software defined receivers. Passive phase shift networks use an array of capacitors, inductors and/or resistors with complexity depending on the accuracy and frequency range required. Active phase shift networks also exist for audio frequencies with op amp ICs being a popular choice.

Phased array: A type of directional antenna whose radiating pattern is determined by the phase relationship (i.e. relative timing) of signals fed to its driven elements. Correct feeding results in a directional beam-like radiation pattern. Phased arrays, with their multiple driven elements, differ from conventional yagis that comprise just one driven element but one or more parasitic elements.

Phased verticals: An HF antenna system comprising two or more vertical elements that have been spaced and fed to provide gain and directivity not possible with a single vertical. Changing phase relationships between the elements allows the 'beam' direction to be switched faster than if relying on a rotary beam. A more advanced project than a simple dipole or vertical, the grounding, spacing and phase differences need to be got right for optimum results.

Phasing harness: A feed arrangement, comprising a coaxial cable T connector and lengths of feedline to allow multiple identical antennas to be connected to the one feedline for increased gain or directivity. Cable in a phasing harness may be adjusted in length to provide a desired impedance transformation and/or phase shift to achieve directivity.

Phasing method: A method of generating single sideband based on splitting audio and radio frequency signals into components that are identical but are 90 degrees shifted in phase relative to one another. Combining allows the unwanted sideband to be nulled out, allowing a single sideband signal to be generated. Can also be used on reverse to receive single sideband signals.

Phone patch: A device used to connect a telephone to a radio transceiver, allowing phone calls to be made via radio. Fitted to many repeaters in the United States (see *autopatch*), they were popular before mobile phones.

Phonetic alphabet: System of assigning words to letters to allow clearer passing of messages under difficult receiving conditions such as weak signals or interference. The International Phonetic Alphabet is the standard for amateur communication but you will encounter alternatives on the air. Sticking to the standard is recommended initially, but you may find cases where an alternative term works better for one or two of your call letters.

Phono plug: Another name for an RCA plug. Commonly used for older style audio and video connections. Their exposed inner pin and risk of short circuits makes RCA connectors unsuitable for use as DC power connections. However, economy-minded ham builders sometimes use them to connect antennas to receiver and low powered equipment. (USA)

Photovoltaic cell: See solar cell.

Pi network: A circuit, containing one inductor and two capacitors, that forms an effective low-pass filter useful for suppressing harmonics. If all are made variable it can be an effective antenna coupler. It gets its name as the arrangement of components resembles the pi symbol (π). Good quality RF low pass filters use multiple pi network sections to obtain a sharper cut-off.

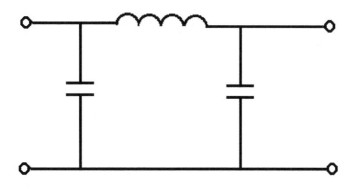

PIC: Peripheral Interface Controller. A group of specialised microcontroller ICs. Can be programmed to perform various electronic, logic and mathematical functions. Commonly used to control electronic equipment.

Pico: $\times 10^{-12}$. One million-millionth. E.g. picofarad (pF).

Picosat: A small, often low earth orbiting, artificial satellite weighing under 1kg.

Piezoelectric: An effect noted with certain types of crystal or ceramic materials where pressure on them generates electrical energy. Or, in reverse, electrical energy generates movement. These characteristics are useful in microphones and headphones respectively.

Pile-up: A large number of stations calling at once, such as to work a rare DX station. Also *dogpile*.

PIN diode: A type of diode use in RF switching applications, especially around transmitter power amplifiers.

Pirate: A person illegally transmitting on a radio frequency. Either they do not have a licence or they breach the terms of the licence they do have. May or may not be a jammer.

Pirate broadcasting: Refers to an unlicensed broadcast station such as may be erected for political activism, particular cultural or musical tastes or fun.

Pitch: Refers to the spacing between connections on a component, plug or socket. For instance, leaded DIP ICs and matrix circuit boards have a pitch of 2.54 mm (0.1 in) between adjacent pins or holes.

PIV: Peak inverse voltage. The maximum voltage a diode can withstand if connected in the reverse biased direction before it breaks down. The PIV voltage of a diode should always be much higher than the voltages normally encountered in the circuit in which it is to be used.

PL: Private Line. See *CTCSS*.

Plate: Old name for the anode connection on a vacuum tube or valve.

Plate modulation: Method of transmitting AM by using the output of an audio amplifier (modulator) to vary the plate (anode) voltage of a vacuum tube final RF amplifier, normally via a modulation transformer.

PLL: Phase locked loop. A circuit using a voltage-controlled oscillator and phase detector in a feedback loop to provide a stable but adjustable frequency such as required in radio transmitters and receivers. PLLs are also used to demodulate FM signals.

PL259: A popular type of coaxial cable connector. Most used on HF and VHF equipment. PL259 is the plug that mates with the SO239 socket. The PL259/SO239 combination is also sometimes referred to as 'UHF connectors'.

PMR446: Personal Mobile Radio. UHF low power licence-free short-range radio communication available in most European countries. Uses channels around 446 MHz. Similar in concept to the Family Radio Service in the USA.

PNP: Relates to a transistor whose emitter is more positive than the collector. PNP transistors are less common than NPN devices but are still used for certain switching applications and, in conjunction with NPN transistors, push-pull audio amplifiers. Opposite to *NPN*.

Polarisation: Refers to the orientation of the electric field lines in radio signals radiated or received by an antenna. Polarisation can be vertical, horizontal or circular. Polarisation's importance depends on the type of wave propagation you are exploiting. For example, on VHF and UHF your antenna's polarisation should match what everyone else is using. That means vertical for FM and repeaters and horizontal for SSB. Whereas for most HF contacts, because signals tumble around in the ionosphere, it matters much less if your antenna is vertical and other people are using horizontal antennas. Also see *cross-polarisation*.

Polarised: Refers to a component that needs to be connected a certain way (or polarity) in a circuit to work. Includes diodes, transistors, integrated circuits and some types of higher value capacitors.

Polyester capacitor: A non-polarised capacitor used in applications where medium values of capacitance are required such as in all types of electronic equipment. Typical values are 1nF to 1 uF.

Polystyrene capacitor: A non-polarised capacitor used in applications where low values of capacitance are required such as RF circuits. Available in high voltage ratings they offer good stability for critical applications. Typical values are 5pF to 1 nF.

Portable: Refers to amateur radio activity away from home, especially outdoors e.g. from a summit, park or beach.

Positive earth: Refers to a wiring system where the positive connection of a battery or power supply is connected to chassis, ground, a metal case or a vehicle's housing. This arrangement is rarely used now because cars generally use negative earth systems and NPN transistors are more commonly used in equipment. Opposite to *negative earth*.

Positive feedback: Refers to an amplifier stage that has had a portion of its output fed to its input to increase its gain and (potentially) cause it to oscillate. Positive feedback is not normally encouraged but can occur due to poor circuit layout or shielding. A regenerative receiver is an example where positive feedback is encouraged to provide higher gain and reception of SSB and CW signals.

Potential difference: Another term for *voltage*.

Potentiometer: A variable resistor with three connections widely used for volume and level controls in equipment by acting as a variable ratio voltage divider. See *variable resistor*.

Power output: The strength of signal emitted by an oscillator, amplifier or transmitter. Measured in watts.

Power amplifier: An amplifier driven by a transmitter to make the radiated signal stronger. See *linear*.

Power budget: A calculation most commonly used to assess whether batteries carried are of sufficient capacity to power portable equipment over a desired period. Information needed includes supply voltage, battery capacity, equipment current consumption and an estimate of *duty cycle*.

Power supply: An item that provides power to a device. Often plugs into a wall socket, converting AC to DC at the required output voltage. Linear types comprise a transformer, rectifier, filter capacitor, regulator and, for high current supplies, pass transistors. Newer switch mode types feature a lighter transformer and switching regulator.

Power transistor: A transistor that can handle high power, such as required for heavy duty switching or amplification. Housed in a large case that can be screwed to a heat sink. Opposite to *small signal transistor*.

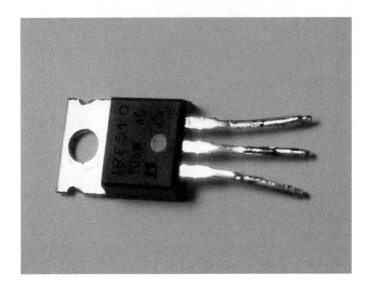

P-P: Peak to Peak. Measurement of an AC signal's level taken between negative and positive peaks such as can be viewed on an oscilloscope. Equal to twice the peak and 2.828 times the RMS amplitude.

Preamp: Preamplifier. Amplifier before a stage or device to boost gain. Examples include microphone preamplifiers to boost transmitted audio or RF preamplifiers to increase receiver sensitivity.

Preferred values: A system of setting standard values for manufactured electronic components so that they are spaced by a roughly similar percentage from the ones above and below. Resistors and capacitors often use the E12 series, i.e. 1.0, 1.2, 1.5, 1.8, 2.2, 2.7, 3.3, 3.9, 4.7, 5.6, 6.8, 8.2 and 10. However a less common E24 series, including the E12 series plus intermediate values also exists, particularly for resistors.

Prefix: The first part of an amateur call sign that identifies its origin country. Some countries also use prefixes to indicate an operator's call area (e.g. region or state) or their licence level. A prefix is followed by a one to four lettered suffix following the number.

Preselector: Tuned circuit in the front end of a receiver adjusted to pass signals on the desired frequency of reception and reject others. Reduces the risk of overload and images from strong *out of band* signals.

Preset: Refers to factory adjustments that once made are left alone. Traditionally internal switches, trimpots and trimcaps, preset adjustments are increasingly menu-based or made via a computer interface or programming cable. Do not touch unless you have appropriate test equipment and know what you are doing.

Primary battery: A battery that can be used only once and not recharged. Examples include zinc-carbon, most alkaline and some types of lithium batteries. Opposite of *secondary battery*.

Primary user: In radio spectrum allocation refers to a user who has priority over other (secondary) users, who must not cause harmful interference to primary users on shared bands. An example (in some countries) is 7.2 to 7.3 MHz, where broadcasters are primary users but amateurs are secondary, meaning that we must not cause interference to broadcasters.

Printed circuit board: See PCB.

Product detector: A circuit stage in a receiver that converts radio to audio frequencies. So called as its output is a product of inputs applied (typically one is the incoming signal and the other is a signal generated within the receiver). Widely used for CW, SSB and digital mode reception.

Propagation: Refers to how signals get from transmitter to receiver. Key modes include ground wave (most notable at lower frequencies) and sky wave (most notable on high frequencies) and direct (or line-of-sight) wave on VHF, UHF and higher frequencies.

Protection diode: A diode, normally inserted at the DC power input of equipment, that prevents damage if polarity is accidentally reversed. Protection diodes can either be wired in series or parallel with equipment. Parallel connection is generally favoured as there is no voltage drop and a series fuse can be installed to quickly remove power.

Prototyping: The process of building, rebuilding and refining electronic circuits until they work satisfactorily. Homebrew prototypes often end up being permanent because the builder is afraid that they won't function as well when tidied up and put in a nice box.

PSE: Please. E.g. PSE RPT is Please repeat, PSE QSL is Please send QSL card, etc. (CW abbreviation)

PSK (or PSK31): Phase shift keying. A narrow bandwidth digital transmission mode that allows keyboard to keyboard chats.

PSU: Power supply unit. See *power supply*.

PTO: Permeability tuned oscillator. A radio frequency oscillator whose frequency is varied by adjusting an inductor.

PTT: Push to talk. The button on a microphone that enables a speech transmitter to transmit. Must be held down while talking.

Q

Q (quality): A measure of the sharpness, efficiency or quality of a tuned circuit, capacitor or inductor. High Q means a sharply tuned circuit with low loss. Think of it like a bell or gong that has been struck. A high Q version would keep ringing well after the strike, with the sound only slowly decaying in volume. Whereas a low Q version would stop ringing quickly. Q is a ratio so is expressed as a number with no unit.

Q (signal): Quadrature. Term for signal that is 90 degrees out of phase relative to an I (in-phase) signal. Often encountered in the design of software defined radios and phasing SSB equipment. Opposite to *I (signal)*.

Q-code: A series of abbreviation codes developed to speed Morse code communication. All have three letters starting with Q. The more common Q-codes are listed below.

QEX magazine: A technically-oriented magazine published by the ARRL. (USA)

QRL?: Is this frequency in use? A courtesy to ask this before transmitting. This Q code is used on CW only; on voice you would simply ask.

QRM: Interference from man-made sources. E.g. another station close by in frequency.

QRN: Interference from natural sources. E.g. lightning static.

QRP: Low transmit power. Often accepted as being as being 5w or under (CW and digital modes) or 10w or under (phone).

QRP ARCI: QRP Amateur Radio Club International. US-based club devoted to low power (QRP) amateur radio communication.

QRPp: Very low transmit power – e.g. 1 watt or less.

QRQ: Sending faster (CW).

QRS: Sending slower (CW).

QRSS: Very slow speed Morse code intended to be received on a computer screen and displayed graphically.

QRSS grabber: An internet-connected receiver that receives and decodes QRSS transmissions. QRSS grabbers may be online so you can see results of QRSS signals that you and others send.

QRZ?: Who is calling me? It is not, despite common misuse, a synonym for CQ.

QRZ.com: A popular amateur website featuring personal profiles, news and discussion forums. http://www.qrz.com

QRZ profile: A page on the qrz.com website that features your personal details, equipment, QSL information etc. Any amateur can write a QRZ profile that tells those they contact a little about them.

QSB: Fading (of signal). Mainly used on CW.

QSD: Bad sending. (Morse code)

QSL?: Did you unambiguously receive the information I just said? If so, reply QSL to confirm. Often used when signals are weak and your contact is not sure you have heard their callsign, name or signal report correctly.

QSL bureau: A centralised means of exchanging paper QSL cards to save on postage costs at the expense of speed. Often maintained by national radio societies. See direct.

QSL card: Originally a postcard confirming an amateur contact has taken place as was required to claim many awards. Now also produced in electronic form to be used in conjunction with faster and cheaper online QSLing methods.

QSL information: Details you need to send a QSL card. May include the address for the QSL manager and any postage costs they would like you to pay.

QSL manager: A person that has been appointed to handle the QSL card paperwork for a station, particularly one that makes many contacts such as a DXpedition.

QSO: A contact or conversation made on amateur radio.

QSO party: A relaxed and leisurely contest. (USA)

QSP: A message for a someone else conveyed by radio.

QST (magazine): The main magazine published by the American Radio Relay League. (USA)

QST (Q-code): A special message for radio amateurs.

QSY: Move to a different frequency.

QTH: Location. Where you are transmitting from.

Quad: A directional gain beam antenna made with elements formed by square loops of wire. The simplest, known as a cubical quad, comprises a quad loop driven element plus either a director or reflector loop.

Quad loop: A single element loop antenna made from wire with four sides of equal length. Typically 1 wavelength perimeter. A quad loop may be used as the driven element in a quad beam antenna. They are often mounted in the vertical plane for higher HF use and horizontal plane (i.e. parallel to ground) on lower HF bands. Such enclosed loops are widely praised for their low noise pick-up.

1 wavelength perimeter

feedpoint

Quagi: Directional beam antenna using both quad and yagi elements. Uses loops for the driven and reflector element and yagi-style straight elements for the directors. Most used on the higher VHF and UHF bands.

Question bank: A selection of questions that may be asked in an examination. Some countries publish their amateur question banks to assist those studying. Others oppose this on the grounds that this encourages 'parrot learning' without understanding key concepts.

R

R: Term used to refer to electrical resistance as in the ohms law formula. May also refer to resistors on a circuit diagram or parts list.

RACES: Radio Amateur Civil Emergency Service. The government arm of amateur radio emergency services run through local civil defence organisations. (USA)

RadCom: Radio Communications. Magazine of the Radio Society of Great Britain.

Radial: A wire extending from the base of a vertical antenna used to form a conductive ground plane and improve radiation efficiency. Radials may be elevated above ground or buried. Buried radials look neater but require more work and a larger number to provide the same efficiency as a smaller number of elevated radials.

Radiation angle: Relates to the angle that a HF signal arrives from the ionosphere. This typically varies with distance, with long distance signals typically arriving at lower angles than signals a few hundred kilometres away.

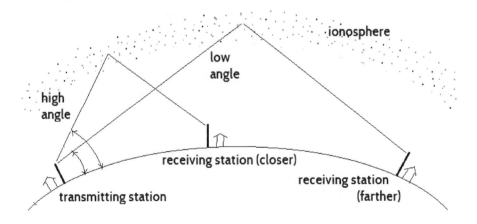

Radiation pattern: Refers to the directionality or otherwise of an antenna when viewed from above. Omnidirectional antennas radiate equally in all directions while directional antennas will have peaks (lobes) and troughs (nulls).

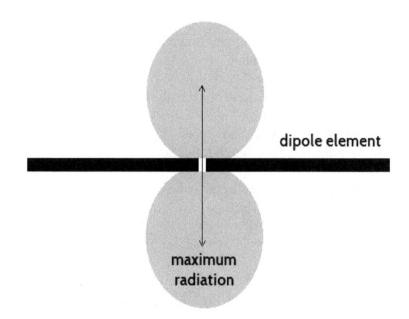

Radiation resistance: The 'good' part of an antenna's feedpoint resistance caused by the radiation of electromagnetic energy from it. Opposite to loss or ohmic resistance. An efficient antenna has a high ratio of radiation resistance to loss resistance. In contrast, antennas such as short helical or loaded vertical mobile antennas or many magnetic loops, have a low ratio and can be very inefficient, particularly on lower frequency HF bands.

Radio astronomy: A facet of astronomy based on receiving radio signals from objects in space.

Radio spectrum: The portion of the electromagnetic spectrum, ranging from very low to extra high frequencies, as used for radio communication. See *RF*.

Radiogram: A paper form for amateur radio message handling. (USA)

Radiosport: Term used in some countries (particularly China and Eastern Europe) for competitive amateur radio activities such as amateur radio direction finding.

Ragchew: A long and leisurely conversation via amateur radio.

Rain scatter: A propagation mode that allows reception of microwave radio signals through reflections from rain clouds.

Rain static: Naturally-caused radio frequency interference sometimes heard when it's raining.

Rally: A hamfest. Organised flea market of new and used amateur equipment. (UK)

Range: Refers to the distance that a radio signal can be heard. Affected by factors such as operating frequency, height and terrain, transmitter power, antenna gain, receiver sensitivity, noise levels and propagation conditions.

Rare DX: Refers to countries that, due to their small number of amateurs, are rarely heard on the air. Sought after by DXers wishing to work as many DXCC entities as possible.

Raspberry Pi: A small microcomputer useful for light-duty tasks. Often incorporated in modern electronic and homebrew radio projects. https://www.raspberrypi.org

RAYNET: Radio Amateurs Emergency Network. Voluntary organisation for radio amateurs involved in emergency communication. https://www.raynet-uk.net/ (UK)

RBN: Reverse Beacon Network. A system of networked receivers that automatically decodes Morse transmissions and display callsigns heard on a website. To use RBN call CQ in Morse and see if your callsign appears on the website. http://www.reversebeacon.net

RC filter: A type of passive filter, using resistors and capacitors only, mainly used to filter audio signals. May be high or low pass depending on the circuit configuration.

RCA connector: A type of plug and socket often used for audio and video connections.

RDF: Radio direction finding (or finder). Techniques or equipment that can identify the direction that a radio signal is coming from. Used for navigation, interference location and *foxhunting*.

Reaction: Another name for the regeneration control in a regenerative receiver. This controls the amount of positive feedback, and, if advanced sufficiently, allows the receiver to oscillate and permit the reception of CW and SSB signals.

Reactive: Refers to a load (such as presented by a non-resonant antenna or a subsequent circuit stage) that presents a high value of either capacitive or inductive resistance at a specified frequency. Power transfer is inefficient unless an impedance transformation circuit, such as an antenna coupler or matching network, is inserted. Reactive loads can also damage components such as RF power transistors, especially if they are used near their maximum ratings.

Readability: The extent to which a signal can be understood. Often expressed as part of an RST signal report, with readability being the first number given. This is on a scale of 1 to 5 with 1 being unreadable, 3 readable with considerable difficulty and 5 being perfectly readable.

Receiver: Electronic device that captures a radio signal via an antenna and converts it to an intelligible signal that can be understood by a person or computer.

Receiving antenna: An antenna only used for receiving signals. This can be optimised to minimise noise pick-up and produce an easier to copy signal than if the transmitting antenna was used. Examples include low wires and rotatable directional loops.

Reciprocal licensing arrangements: Arrangements between countries that allow you to transmit for a limited time when travelling without having to sit a local examination. See *CEPT* and *IARP*.

Reciprocity: The principle that what happens in one direction also happens in the other direction. For example, a beam antenna that exhibits a gain and directionality advantage relative to a dipole on transmit should exhibit similar characteristics on receive. In practice other factors may complicate things, for example a small antenna may be very lossy on transmit but perform well on receive due to its ability to minimise interference pick up.

Rectification: The action of a diode, which conducts current in one direction only. Uses include converting AC signals to DC, converting RF signals to audio or generating harmonics. Rectification can occur naturally, for example when dissimilar metals are in contact with one another in the presence of a strong radio signal. This is why masts and gutters near antennas need to be kept in good condition free from rust to prevent harmonic radiation and reception.

Rectifier: Another name for a diode. Commonly used to convert AC to DC in power supply circuits, detect RF signals and perform frequency conversions in mixer circuits.

Reduction drive: A gear system attached to a variable capacitor to make tuning a receiver, antenna coupler or magnetic loop antenna easier. Also referred to as 'vernier reduction drive' or 'slow motion drive'.

Reference: An item of known size, performance or characteristics used to calibrate, compare or test equipment. For example a frequency reference may be used to calibrate equipment, or a known reference antenna may be used as a baseline against which to compare the performance of a new or experimental antenna.

Reflected power: RF power that is reflected (or returned) up a transmission line due to an impedance mismatch between the transmitter and the antenna. Reflected power is indicated on a reflectometer or VSWR meter. High amounts of reflected power, caused by the use of a non-resonant antenna, indicates an impedance mismatch. This further increases loss and lessens your station's efficiency, especially if a coaxial feedline is used. Opposite to *forward power*.

Reflector (antenna): The rearmost element in a quad or yagi beam antenna. Mounted behind and slightly longer than the driven element.

Reflector (internet): An old name for an email-based discussion list. Many of these exist for various facets of amateur radio.

Regeneration: Another name for positive feedback. Normally undesirable, it can be useful in simple types of receivers to increase gain and/or allow SSB and Morse signals to be received.

Regenerative receiver: A simple form of radio receiver that uses positive feedback (or regeneration) to increase the amplifying capacity of a tube, valve or transistor. Used in the early days of radio, regenerative receivers need few parts to receive signals. Able to receive AM, CW and SSB signals they still make a good project for someone who wants to make a simple receiver. Limitations include frequency drift and poor selectivity, especially against strong signals on nearby frequencies.

Relay (communications): Refers to the passing of messages via an intermediate station, especially between operators unable to hear one another directly. The intermediate station can either be a manual operator or automated device such as a VHF or UHF repeater.

Relay (component): An electromagnetic component containing an electromagnet that controls switch contacts when energised. Useful for safely switching high current from a low current or switching an antenna between the transmitter and receiver stages in a transceiver.

Remote base: See *remote station*.

Remote head: Refers to a front control and display panel of a mobile transceiver that is detached from the main transceiver but connected to it via a cable. Separation allows the panel to be mounted in a more convenient location while the main transceiver unit can be positioned where there is more space.

Remote receiver: Refers to a receiver in an alternative location that may be higher, better located or suffer less interference. The signal from the receiver to the operating location is then sent via an internet or radio connection.

Remote station: An amateur radio station, often in a favourable location, that can be operated remotely via an internet link. Remote stations are used by amateurs who wish to operate but can't set up a station at home. For example, an apartment dwelling ham with a country property may erect a private remote station for their own use. Other remote stations are built by radio clubs for members or operate as a fee for air time commercial service.

Repeater: A radio installation, typically on a tall building or mountain, that receives and retransmits signals from low power, mobile, handheld or poorly located stations to allow communication over an extended range. Repeaters are most prevalent on 2m and 70cm with some on other bands like 10m, 6m and 23cm. Repeaters listen on one frequency and transmit on another in the same band, with the separation called an offset or split. They are typically erected and maintained by local radio clubs.

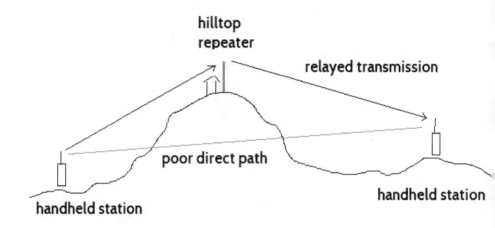

Repeater linking: The practice of connecting two or more radio repeaters together to enable communication beyond the range of one repeater alone. Linking may be done via radio, a telephone line or the internet.

Repeater sponsor: The individual or club that hosts and maintains a repeater.

Resistance: The extent to which a component retards current flow. Measured in ohm.

Resistor: An electronic component that limits the flow of current. Applications include transistor biasing, signal attenuation, dividing voltages and more. Values are measured in ohm, kilohm (1000 ohm) or megohm (1 000 000 ohm). See also *variable resistor*.

Resolve (a signal): To adjust a receiver so that an incoming signal can be understood or decoded.

Resonant antenna: An antenna that presents a purely resistive load to the transceiver at a desired frequency. If this load is close to 50 ohm it can be directly connected to a transceiver with no transformer or antenna coupler required. Common examples of such resonant antennas include half wavelength dipoles, quarter wavelength ground planes and most beams. Opposite to *non-resonant antenna*.

Resonant frequency: The frequency at which an antenna acts as a purely resistive load with no reactive components. Capacitive or inductive reactance appears and then increases as the applied frequency is moved off resonance.

Restrictive covenant: A petty regulation, often specified by a body corporate or homeowner association, that limits what a home owner or resident can do to or around 'their' home. For example, restrictive covenants may prohibit people from painting their fence purple, keeping multiple pets or erecting visible antenna masts. Also known as CC&Rs, they can be avoided by purchasing a home not in an estate featuring these rules.

Rettysnitch: A sharp probe-like instrument that according to ARRL lore was supposed to enforce decency on the amateur airwaves. See *Wouff Hong*. (USA)

Return loss: The loss of power in an RF signal reflected by a mismatched RF transmission line. Inversely related to standing wave ratio but expressed in dB. Unlike other types of loss, where high numbers are bad, a low return loss indicates a poor match while a high return loss indicates a good match.

Reverse: Relates to frequencies opposite to which you are transmitting and receiving on when using an FM repeater. Listening 'on reverse' means listening to the frequency stations use to access a repeater. In other words, the input frequency. This is useful to determine whether they are within range and communication can be established directly (or simplex) without the repeater.

Reverse polarity: Refers to transposing DC power connections so that an electronic item's positive lead goes to the supply's negative connection and the negative lead goes to the positive contact. This endangers equipment not fitted with polarity protection circuitry. Good construction practices minimise the chance of reverse polarity by employing polarised connectors. Within circuits components like diodes and electrolytic capacitors can be connected reverse polarity. The effect is (at best) the item not working, or (at worst) component or equipment damage.

RF: Radio frequency. Refers to frequencies that occupy the spectrum above sound but below light. Part of the electromagnetic spectrum, radio frequencies are divided into segments e.g. very low frequency (VLF), low frequency (LF), medium frequency (MF), high frequency (HF), very high frequency (VHF), ultra-high frequency (UHF) and super high frequency (SHF), each with differing propagation characteristics, antennas and equipment construction techniques.

RF ammeter: Radio frequency ammeter. Item of test equipment used to measure antenna current. Useful when adjusting antenna and ground systems, especially on MF and lower HF bands. Older RF ammeters used a heated wire or thermocouple, which could fail if overheated. Newer instruments use a transformer in series with the antenna connection.

RF attenuator: Radio frequency attenuator. A control, switch or piece of equipment used to reduce the strength of a radio signal, such as required on a receiver, etc to prevent over-load. Also see *step attenuator*.

RF burn: Radio frequency burn. An injury caused by bodily exposure to high RF voltages. RF burns can cause scars that never heal. Preventative measures include not touching antennas while RF energy is being applied and mounting them high above human reach.

RFC: RF choke. Radio frequency choke. An inductor used to isolate radio frequency signals from unwanted parts of a circuit, such as audio and power connections. RF chokes will pass DC but block high frequency AC signals.

RF feedback: Radio frequency feedback. A problem where radio frequencies re-enter an earlier part of the transmitter chain and cause distorted transmissions. Can be caused by poor equipment design, lack of shielding, feedline radiation or an antenna that is too close to the transceiver.

RF fingerprint: Radio frequency fingerprint. Relates to certain subtle characteristics of a transmitted signal that, with the right equipment, can be used to identify a transmitter being used to cause interference.

RF gain: Radio frequency gain. A control on a receiver that allows you to lower its sensitivity. This may be required if a very strong signal is overloading your receiver and causing reception to be distorted. Also *attenuator*.

RFI: Radio frequency interference. Refers to the emanation of radio frequency noise from an electrical or electronic item such that it interferes with the reception of signals on a nearby receiver or transceiver. The interference may either be radiated through the air or conveyed through a common power connection. Normally caused by shoddy design or construction such as poor shielding or the absence of RF bypass capacitors.

RF noise: Radio frequency noise. See ambient noise.

RF preamp: Radio frequency preamplifier. An amplifier circuit connected between the antenna and a receiver's antenna socket to improve sensitivity. Most used on VHF and UHF. Also *masthead amplifier*.

RF probe: Radio frequency probe. An attachment to a multimeter that allows you to measure RF voltages in transmitter and receiver circuitry.

RF quiet: Refers to a location with little natural or man-made RF noise. Very suitable for receiving weak signals that would not be heard elsewhere.

RF speech processor: A type of audio speech processor that increases the average volume of transmitted speech by converting audio from the microphone to a radio frequency, clipping it, filtering it and then mixing it back down to audio. This approach, which allows more compression before distortion is introduced, is commonly used by SSB DXers who need a punchy penetrating signal.

Rheostat: A type of variable resistor with only two connections. Can be used to vary the current that can flow in to a circuit.

Rhombic: A large four-sided wire antenna used for HF transmitting and receiving over a wide range of frequencies. Its plane is parallel to the ground. Termination of one end with a resistor makes its radiation unidirectional.

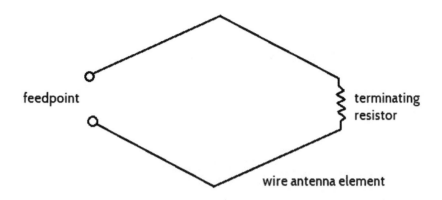

Ribbon: A type of antenna feedline comprising two parallel wires separated by plastic. At one time popular for TV antennas. The slotted ribbon variety is lighter and lower loss so is preferred for amateur use. Ribbon feedline has a characteristic impedance of 300 ohm so cannot be directly connected to a transceiver without a matching transformer, balun or balanced antenna coupler in between.

Rig: A transmitter or transceiver.

Ripple: Small cyclical voltage variations or 'noise' on what should be a steady DC voltage. Caused by poor filtering within a power supply or stray coupling from nearby AC noise sources.

RIT: Receiver Incremental Tuning. A control that allows a transceiver's receive frequency to be varied without changing the transmit frequency. This is useful if on a net and one station is off frequency. Also known as *clarifier* or *fine tuning*.

Rocking armature: type of dynamic microphone or earpiece as widely used in radio transmitters and telephone receivers.

Roger beep: A beep that some have on the end of their voice transmission to inform others that their transmission has finished. Sometimes used as a novelty by CBers, its more serious use is under weak signal conditions where the beep cuts through better than speech.

Roller inductor: A variable inductor whose value is continuously adjustable with a roller contact assembly. Used in high quality antenna couplers and matching networks.

Rotary encoder: A type of continuously turning rotary switch that produces special outputs when turned such as is often required for DDS VFOs or menu adjustments on modern equipment.

Rotary switch: A switch that requires a turning action to select between positions. Rotary switches usually have multiple switch positions (unlike most toggle switches). This makes them useful in antenna couplers to adjust inductance. Rotary switches were also popular in older receivers and transmitters to switch band and mode.

Rotator: A motor and gear assembly mounted on a mast that allows a directional antenna to be pointed in a desired direction. Typically adjusted in the shack with a manual rotator control or automated with computer software. Rotators for HF and terrestrial

VHF/UHF antennas rotate for azimuth only while those for satellite and EME stations also change elevation (az-el rotators).

Rotator controller: A device in the shack that allows you to control a beam antenna rotator and set its direction.

Rotor: Refers to the spindle and moving vanes of a variable capacitor. Often connected to the capacitor's frame. Opposite to *stator*.

Round: A division of time in a group contact or net marked by each participant having one turn to transmit. Imagine a net where Station 1 is the controller and stations 2, 3 and 4 take it in turns to talk. The end of Station 4's transmission marks the end of the first round. The net controller may say a few words, call for further check-ins and commence the second round, with stations taking it in turns to transmit. Those wishing to leave the net soon but not immediately might announce that they will "go one more round" before signing.

RPT: Repeat. Or (signal) report. (CW abbreviation)

RSGB: Radio Society of Great Britain. The representative society for radio amateurs in the United Kingdom. http://www.rsgb.org

RST: Readability Strength Tone. Common method of giving signal reports. R is readability on a scale of 5 (0 unreadable, 5 fully readable with no difficulty). S is signal strength on a scale of 9 (0 is no signal, 9 is very strong signal). T is tone on a scale of 9 (0 is rough tone, 9 is perfect tone) used for CW only.

RTL-SDR: A popular plug-in USB dongle that can be used to receive radio signals with a computer and suitable software. A low-cost way of receiving HF, VHF and UHF radio signals. https://www.rtl-sdr.com/

RTTY: Radioteletype. An originally mechanical form of sending text over radio through the use of an electric typewriter-like machine called a teleprinter. RTTY signals can now be generated and received with computers.

Rubber ducky: The short plastic-covered whip antenna often supplied with portable transceivers. Rugged but offers poorer performance than a full-sized antenna.

Running: Operating with, especially with reference to transmitter output power. E.g. "running 100 watts into a dipole".

Rx: Receiver or receiving. Opposite to *Tx*.

S

S-meter: Signal strength meter. In HF receivers and transceivers to give an indication of incoming signal strength. S0 is no signal, S1 is very weak, S9 is strong, while S9 +40dB is extremely strong.

Sangean: Manufacturer of popular portable shortwave receivers. http://www.sangean.com

SAREX: Shuttle Amateur Radio Experiment. A program of experiments done by crew of the Space Shuttle involving amateur radio. Now replaced by *ARISS*.

Scan: A function found mainly on VHF/UHF equipment that automatically scans through stored memory channels or over a frequency range and stops on occupied frequencies. This allows activity to be monitored without manually tuning a receiver.

Scanner: A VHF/UHF receiver capable of receiving local amateur, airband and other communication. So called due to their ability to hunt for activity by scanning memory channels. Radio scanning used to be a popular hobby but declined after many spectrum users switched to more secure digital radio systems.

Schematic diagram: A diagram using abstract symbols for electronic components to more easily convey a circuit's function. Also *circuit diagram*.

Schottky diode: A type of diode able to switch quickly, as useful in RF circuits. Also *hot carrier diode*.

Scope: See *oscilloscope*.

SCR: Silicon Controlled Rectifier. A current controlling component as often used in power supply switching and protection circuitry. It works by conducting when a small voltage is applied to its gate terminal.

Scramble: A short contest with simple rules. Often locally-organised. Similar to a *sprint*.

Screwdriver expert: A person who doesn't know what they don't know about servicing radio equipment. Prone to tweak internal adjustments to increase a transmitter's RF output power but in doing so may damage a slug and make performance worse. 1970s expression popularised by a top-selling CB radio book.

SDR: Software defined radio. An approach to radio transmission and reception where signal generation and processing is done through computer software instead of electronic hardware.

SDR dongle: An accessory that, in conjunction with SDR software, allows a computer to operate as an SDR receiver. Normally plugs into a USB socket. Most cover VHF and UHF frequencies only but can receive HF with an *upconverter*.

SDR Sharp (SDR#): Popular software that allows a software defined radio dongle to operate. https://www.rtl-sdr.com/sdrsharp-users-guide/

Search and pounce: A method of obtaining contacts by tuning the band and either replying to other station's CQ calls or tail-ending. Particularly effective with modest antennas or if *QRP*.

Secondary battery: A battery that can be recharged many times. Opposite to a single-use *primary battery*.

Secondary user: In radio spectrum allocation refers to a user who must not cause harmful interference to primary users who have priority on shared bands. An example (in some countries) is 7.2 to 7.3 MHz, where broadcasters are primary users but amateurs are secondary, meaning that we must not cause interference to broadcasters.

Selcall: Selective calling. A technique where a special series of tones from a transmitter unmutes only a targeted receiver or group of receivers. Most used on commercial VHF/UHF communication systems.

Selectivity: The ability of a receiver to reject unwanted signals on adjacent frequencies. Determined by the quality of its filters.

Self-amalgamating tape: A specialised type of tape that when applied forms a protective insulating waterproof mass around the item it wraps. Commonly used to protect outdoor electrical and antenna connections from corrosion.

Self-oscillation: Refers to a malfunctioning amplifier stage acting like an oscillator and generating spurii.

Self-regulation: Concept that amateurs are technically trained and responsible people who can regulate their own affairs with minimal involvement of (often underfunded) government regulatory agencies.

Self-resonant (frequency): Refers to the frequency that an inductor resonates at in conjunction with its stray capacitance.

Self-spotting: The act of reporting yourself on a DX cluster or other online forum in order to drum up contacts. This is acceptable for awards like SOTA but often not for contests whose rules may prohibit this.

Semi break-in: Refers to a feature on a Morse code transceiver that allows automatic switching between receiving and transmitting. Fast switching, known as full break-in, allows listening between letters sent.

Semiconductor (component): One of a family of silicon or (rarely) germanium-based components such as diodes, transistors and ICs. Often described as *solid-state* electronics.

Semiconductor (material): A substance that can conduct electrical current under some conditions but not others. Examples include germanium and silicon.

Sensitivity: The ability of a receiver to detect weak signals. Determined by its gain and internally generated noise.

Separation cable (or kit): A cable that allows a mobile transceiver's front panel to be detached from its main body. The front panel (or head unit) can be mounted in a spot convenient to the driver while the rest of the transceiver can be mounted elsewhere in the vehicle where there is more room.

Serial number: A number given as part of a contest exchange. In many contests you start with 001 for your first contact, going up by one thereafter. In turn you need to receive, confirm and log the number your contact gives you.

Series: Connecting components in a line such that there is only one possible path for current to take. Opposite to *parallel*.

Series resonant: Relates to a tuned circuit with the capacitor and inductor connected in series. In this arrangement the impedance (i.e. resistance to radio frequencies) falls to a minimum at the resonant frequency. This has many uses in radio circuits. For example, if you have a receiver being overloaded by a strong signal from a nearby radio station you could connect a series resonant tuned circuit across its antenna connection to suppress the unwanted signal but allow signals on desired frequencies to pass unaffected. Also see *parallel resonant.*

SFI: Solar flux index: Basic measure of solar activity as determined by the radio noise emitted by the sun at 2800 MHz (hence 10.7cm solar flux). Closely correlated with sunspot numbers and ionisation of the upper ionosphere (F-region). Expressed in solar flux units, 66-70 is considered very low, as experienced at the bottom of the solar cycle, about 100 – 120 is medium and over 200 high, as experienced at the top of the cycle.

Shack: Radio shack. The place where you transmit from. Can be an outside shed, spare room or even cupboard that is closed when not operating.

Shaft: The turning spindle of a control such as a variable resistor, variable capacitor or rotary switch. Normally extended through a front panel with a knob affixed.

Shape factor: The 'tightness' of the band pass characteristics of a crystal or mechanical filter. Expressed as a ratio of the wider bandwidth relative to the narrower bandwidth with the attenuation figures quoted. For example, a crystal filter with bandwidths of 6 kHz (at 60 dB below its peak) and 3 kHz (at 6 dB below its peak) would have a shape factor of 2:1. A smaller ratio means a tighter, more selective filter better able to reject off-frequency interference.

SHF: Super high frequency. The portion of the frequency spectrum between 3 GHz and 30 GHz. Useful for satellite and broadband data communication. Amateurs also have SHF allocations and have spanned impressively long distances.

Shielded cable: A type of cable comprising one or more inner conductors and braided or foil shielding. Examples include cables handling low level audio signals (e.g. microphone cables) or RF energy (such as coaxial cable antenna feedlines).

Shielding: Enclosing sensitive parts of radio circuits in metal cases to lessen interference to their operation or radiation from them. Missing or poor shielding can cause frequency drift and unwanted feedback in radio circuits.

Short circuit: An unwanted electrical connection between parts of a circuit that should be separate. Can be caused by bad wiring, poor soldering or metal debris across components.

Short-form kit: Refers to a kit that may include the circuit board, instructions and basic components. You may need to find other parts such as plugs, sockets, knobs and enclosures.

Short path: Refers to the shortest path around the globe to a distant station. Opposite to *long path*.

Short skip: Refers to radio propagation that permits skywave communication over relatively short distances, e.g. a few hundred kilometres. This is routine on the lower HF frequencies but can also occur on higher HF and lower VHF frequencies due to sporadic-E propagation.

Shorted turn: A defect in an inductor, transformer or balun coil where adjacent windings that should be insulated from one another come into contact due to a breakdown or degradation of the insulation between them. Shorted turns (at best) impair performance or (at worst) cause sparks to fly.

Shortwave: An alternative term for HF or high frequency. That is frequencies between 3 and 30 MHz. Shortwave frequencies are most known for their ability to support long distance and international communications due to their ability to be reflected by the ionosphere.

Shunt: A low value resistor wired across a circuit or device (such as a meter movement) to divert most of the current applied around it. An example would be resistance wire across a meter movement that allows it to read higher currents or transmit powers.

Shunt fed: Refers to a tower that is fed part way up to form a vertical antenna. Most used on the 1.8 – 7 MHz amateur bands. The 14-28 MHz beam antenna normally on top contributes to the vertical's performance by forming a capacitance hat.

Shunt resistor: A low value resistor (which can be a coil of wire) connected across a meter movement when using it to measure current draw. By diverting most of the current around it the meter is made less sensitive. This allows the meter to read full scale when drawing several amps instead of microamps or milliamps as would be the case without the shunt.

Sideband: Refers to each of the ranges of frequencies either side of a carrier signal that carry intelligence. In some cases both sidebands may be present (e.g. double sideband/AM) while in others only one sideband is transmitted (single sideband/SSB).

Sidetone: In a Morse code transceiver an audio oscillator that provides an audio tone in the speaker when the key is pressed. Hearing sent Morse makes sending easier for many people.

Signal generator: A piece of test equipment that generates audio or radio frequency signals for testing amplifiers and receivers etc. Can be varied in frequency and output level for sensitivity tests.

Signal injector: A simple piece of test equipment used to generate a signal for testing amplifier and receiver circuits. The signal injected can be AF or RF, depending on application. Unlike a signal generator, whose frequency can be varied, a signal injector normally operates on one frequency only. Often used in conjunction with a signal tracer, RF probe or oscilloscope to trace what happens to the signal through various stages.

Signal report: A report of how well you receive other stations. Given early in a radio contact after callsigns have been exchanged and confirmed as correct. Signal reports are often required as part of contest exchanges. Signal reports for amateur radio use the RST (Readability, Strength, Tone) scale for Morse code and just readability and strength for voice modes. Computer programs that decode digital modes often give a signal to noise ratio figure expressed in decibels. Broadcast and shortwave listeners have other signal reporting systems, for example the SINPO code.

Signal strength meter: Meter that indicates the strength of an incoming signal on a receiver. Also S-meter.

Signal tracer: A piece of test equipment comprising an audio amplifier used to trace the progress of a signal through an amplifier stage and thus assist in finding faults. Often used in conjunction with a signal injector.

Signed: Signed off. Left the conversation. Refers to a participant in a contact or net who has had their final transmission for the session (e.g. "<callsign> having signed").

SIGS: Signals. (CW abbreviation)

Silent key: A deceased amateur.

Silicon chip: An *integrated circuit*.

Silicon diode: A common type of diode commonly used to convert AC to DC in power supply circuits.

Simplex: Communication where all transmitting and receiving takes place on the one frequency. Typical for most HF contacts and VHF/UHF contacts that do not involve repeater, satellite or cross band operating.

SINAD ratio: SIgnal-to-Noise And Distortion ratio. A measure of signal quality expressed in decibels for a given input signal level. SINAD figures are often found as part of VHF or UHF FM receiver sensitivity specifications. For example a VHF receiver may have a sensitivity of 0.3 uV for 12 dB SINAD. This means that an 0.3 microvolt level signal is required to cause a received signal that is 12 dB above the noise floor. Such a receiver will be more sensitive than one measuring 0.4 uV for 12 dB SINAD but less sensitive than one requiring only a 0.2 uV input for 12 dB SINAD.

Sine wave: A waveform of a pure audio or radio frequency signal that is all one frequency, i.e. it has no harmonics or spurii. Such a signal gives a sine curve shape when its waveform is viewed on an oscilloscope.

Single balanced mixer: A type of RF mixer that suppresses one of the two input signals at the output. This arrangement is used in circuit applications where use of the more complex double balanced mixer, which suppresses both input signals, is not justified.

Single conversion: Relates to the simplest form of superhet receiver in which incoming signals have one radio frequency conversion to one intermediate frequency before being converted to audio. See *dual conversion* and *triple conversion*.

Single hop: Relates to a signal that has needed only one hop via the ionosphere to get from the transmitting to the receiving location. The reduced path attenuation often makes single hop signals very strong, even from stations running modest power and antennas.

Single operator: Refers to a contest or DXpedition station where one person makes all the contacts. Opposite to *multi operator.*

Single signal reception: Refers to a receiver that is sufficiently selective to hear signals on one setting of the dial only. This was difficult in the early days of radio but is routine today with more advanced superhet and phasing direct conversion receivers.

SK: Sent at end of final transmission to signify end of contact. (CW abbreviation)

Sked: Radio schedule. A planned contact with a particular person on a particular frequency at a particular time. E.g. amateurs travelling often set up skeds to contact friends back home each evening.

Skin effect: Tendency of radio frequency signals to flow along the outside of a wire or conductor.

Skip: Refers to HF radio signals being reflected by the ionosphere, 'skipping' over distances between where the ground wave peters out and the signal returns to ground.

Skip distance: Distance between an HF signal's origin and where it first returns to earth. Affected by factors such as frequency, time of day, latitude and solar conditions.

Skip zone: Area over which an HF signal skips when it is reflected by the ionosphere.

Skirt: Refers to the passband shape of an RF band pass filter such as used in a receiver for front-end or intermediate frequency filtering. A filter with a wide skirt offers poorer rejection of off-frequency signals compared to a filter with a narrow skirt. Also see *shape factor.*

Sky wave: Refers to propagation of radio signals via reflection off the ionosphere as frequently occurs to HF signals. Contrasts with ground wave (mainly LF and MF) and direct wave (mainly VHF and UHF).

SLA: Sealed lead acid. A type of rechargeable battery ideal for powering amateur radio equipment in the field. Requires a constant voltage charger. Low cost but heavy. Being replaced by lighter lithium-based batteries.

Slash (/): Sometimes used to identify a special characteristic of an amateur station, such as if you are operating in a different country, mobile or portable. Most often seen in logbook entries or heard on Morse (-..-.).

SLF: Super low frequency. Frequencies between 30 Hz and 300 Hz.

Slide switch: A type of switch that requires a sliding movement to move the contacts.

Slim Jim: A type of vertical antenna commonly used on the VHF and UHF bands. Often made from ribbon or slotted antenna feedline.

Slope detection: A crude technique of resolving a narrow band FM signal on an AM receiver by tuning the receiver off to one side of the signal. While signals heard this way are intelligible, there is some distortion and there is no noise reduction that you would have on a proper FM receiver.

Sloper: A slanted dipole antenna supported by a tower and often comprising one of its guys. The tower's main role is often to support a 14 – 28 MHz beam with slopers providing low-cost capability on 7 and/or 10 MHz.

Slot: Antenna comprising a metal plate punched with holes or slots. Most often used at UHF and higher frequencies.

Slow blow: Refers to a type of fuse that can pass current beyond its rating for some time before it will blow.

Slug: A threaded piece of ferrite or iron powder inserted in a former used to vary an inductance such as required to align a tuned circuit in a receiver or transmitter.

SMA: A small threaded plug and socket combination often used on the antenna connections of handheld transceivers.

Small signal transistor: A type of transistor only capable of handling small signals such as encountered in audio amplifier, receiver and low-level transmitter stages. Opposite to *power transistor*.

Smart charger: Type of battery charger required for more modern battery types, such as lithium ion and lithium polymer. While charging these monitor the condition of the battery, adjusting voltage and current to suit. This ensures optimum charging and prevents damage.

SMD: Surface mount device. See SMT.

Smith chart: A circular chart used to graphically solve problems involving complex numbers such as the reactive impedances often encountered in transmission line and antenna work.

SMPS: Switch mode power supply. See *switch mode.*

SMT: Surface mount technology. Refers to chip-type components that can be mounted on the surface of an etched printed circuit board with no hole drilling required.

Sniffer: A hand-held receiver used to find the exact location of a fox during a foxhunt. May be mounted on the boom of a directional beam or loop antenna. Sniffers feature adjustable sensitivity and an audio or visual indicator of signal strength to allow precise location of a signal's source.

SN: Soon. (CW abbreviation)

SNR: Signal to noise ratio. A ratio, often expressed in decibels, comparing the strength of an incoming signal with the strength of noise in an amplifier or receiver. A high signal to noise ratio means that the signal is much stronger than noise and vice versa. Computer programs that decode digital modes often give an SNR reading for signals detected as a substitute for the RS(T) reports given by manual voice and Morse code operators. Their sensitivity is such that signals below the noise level (i.e. a negative SNR value) can still often be decoded.

SOC: Second Class Operators Club. A Morse code operators club "because so few are really first class". www.qsl.net/soc/

SO239: A popular antenna connection socket found on HF, VHF and sometimes UHF transceivers. Accepts the *PL259 plug*.

Solar cell: A cell that converts light to electricity. Also *photovoltaic cell*.

Solar cycle: See *sunspot cycle*.

Solar flare: An eruption of electromagnetic energy from the sun's surface. The extra radiation ionises the ionosphere's D-layer, causing it to absorb radio signals. This can result in an HF *black out* that can last up to several hours.

Solar indices: Numbers used to indicate solar activity such as is useful to predict ionospheric conditions as they affect radio signal propagation. Examples include the SFI (solar flux index), A-index and K-index.

Solar panel: A panel of solar cells that generates electricity from light using the photovoltaic principle.

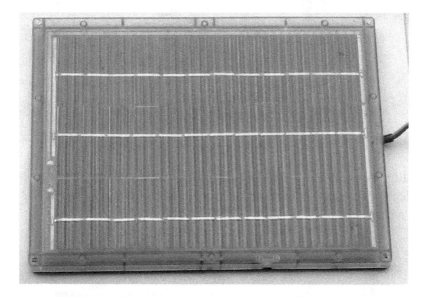

Solar regulator: A voltage regulator that converts the sometimes-varying output from a solar panel to a constant voltage and/or current such as required to charge a battery or operate equipment. Also known as a charge controller, solar regulators often use switch mode regulators that if poorly built can cause radio interference.

Solder: low melting point alloy that when heated bonds wires to create electrical connection. Leaded solder for electronics typically comprises 40% lead and 60% tin with a rosin core to assist bonding. Lead-free solder contains traces of other metals.

Solder sucker: Hand or electrically driven pump that sucks molten solder such as may be required when removing parts from a circuit board etc.

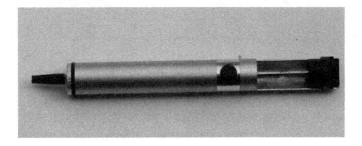

Solder wick: A copper braid, supplied on a roll, that absorbs solder through capillary action. Used to remove solder from a joint such as when removing a component from a circuit board. Also known as desoldering braid.

Solenoid coil: A style of inductor with turns of wire wound parallel to one another in a spiral. The turns may be air-wound or supported by an insulated former. A ferromagnetic slug or rod may be inserted to increase the inductance. Used in many electrical and electronic applications.

Solid dielectric: Refers to a variable capacitor that has a solid dielectric (e.g. plastic or mica) between is plates as opposed to air. This increases the capacitance obtainable from a physically small component.

Solid state: Refers to equipment with transistors and integrated circuits only – no tubes (or valves).

SOTA: Summits On The Air. An awards program that encourages amateurs to make contacts from hill peaks that exceed a specified prominence. https://www.sota.org.uk/

Sound card: An accessory that provides the interface between a computer and audio inputs and outputs. Sound cards are important for amateur radio applications including digital modes and software defined receivers. Computers normally have internal sound cards but an external low noise stereo sound card can provide improved performance such as desirable for software defined radios.

Sound card mode: A mode that requires a connection between a transceiver and a computer via its sound card in conjunction with a special program. Previous electro-mechanical and electronic modes such as RTTY and SSTV that required dedicated equipment have become sound card modes. More modern sound card modes include WSPR, PSK31, Olivia, FT8 and other digital modes.

Spacer: A tubular piece of plastic or metal often with a bolt passing through used to hold a circuit board, assembly, display or panel inside an enclosure.

Spark transmission: The original transmitting mode for wireless telegraphy before CW (Continuous Wave). Generated by using high voltages to create a spark and release electromagnetic energy. Spark transmitters used a tuned circuit to concentrate its signal around a particular frequency and couple its output to an antenna.

Speaker-microphone: Typically plugged into a handheld transceiver, this accessory replaces its internal speaker and microphone. An external speaker-microphone allows the transceiver to be clipped to a belt or held higher for a better signal. Audio quality may also be improved.

Special event station: A station established by a radio club in a public location to promote and support a locally or nationally significant event or anniversary.

Special event callsign: A temporary callsign obtained for a special event station. May include letters that abbreviate or spell out the name of the event.

Spectrum analyser: Piece of electronic test equipment that allows you to monitor signals on a frequency scale. This allows you to observe occupied bandwidth and any spurious signals.

Spectrum display: The main visual display on a spectrum analyser or receiver. Frequency is normally along the X-axis while amplitude (level) is normally shown on the Y-axis. Such a display is useful to show the desired signal's bandwidth, the power distribution within it and any spurious sidebands. Alternatively, if broadened, the spectrum display can show all activity on a band or band segment. This is handy to visually gauge propagation and activity.

Speech amplifier: See *microphone amplifier*.

Speech clipper: An electronic circuit, often comprising diodes or transistors, that removes the peaks of audio signals. This can be beneficial to prevent over-modulation and allow the average level of a signal to be raised for improved readability.

Speech compressor: See *audio compressor*.

Speech processor: A circuit that boosts quieter parts of your voice to increase the average loudness of your transmission and improve intelligibility if signals are weak. Can cause distortion or splatter if incorrectly adjusted. While useful on SSB, speech processors should be switched off if transmitting digital modes or SSTV. Also see *RF speech processor*.

Splatter: Transmission of signals outside the normal bandwidth of a transmission due to excessive microphone gain, over-driving, etc. The result is distorted transmissions and potential interference to others.

Split: See *offset*.

Split frequency: Refers to a contact where you are transmitting on a different frequency (but normally within the same band) to the station you are working. This is commonly done by sought after DX stations who have so many stations calling that it is hard for them to pick out one to contact. The use of split frequency spreads out calling stations and allows the DX station to more easily select one to work.

Sponsor: Person or club who maintains a communications facility such as an amateur radio beacon, repeater, or remote station.

Spot: Indicates that your signal has been heard and reported on the DX cluster or other online forum. This may encourage others to call you. Also see *self-spotting*.

Spreader: Insulated pole or pipe used to keep antenna wires separate, such as may be used in a fan dipole, cubical quad or Moxon antenna.

Spread spectrum: A wide bandwidth mode that uses a broad range of frequencies. Particularly used for defence communication due to its robustness against interference and jamming.

Sprint: A short contest. It may only go for an hour on a single band. Similar to scramble.

Sporadic-E: A means of radio signal propagation that allows extended distance communication on the 28, 50 and 144 MHz amateur bands due to reflection from the ionosphere's E layer. 500 to 1500 km distances are most common. Sporadic-E can occur at any phase of the sunspot cycle and is most prevalent during mid-summer.

Spotting: Refers to alerting others, for instance through a DX cluster, forum or social media, that a particular station (often rare or long distance) is on the air.

SPST: Single Pole Single Throw. The most basic type of switch with two contacts that are either open or closed circuit.

SPDT: Single Pole Double Throw. A type of switch where the middle (common) contact can be switched between two outer contacts. One of the outer contacts is not used if using it as a simple power switch.

Spurii: A stray or spurious signal such as generated from a badly built, shielded or adjusted transmitter. Can cause interference to other spectrum users.

Square wave: A signal whose level changes abruptly between minimum and maximum, with equal time spent at each and no time spent at intermediate values. Unlike pure sine waves, square wave signals are rich in harmonics and need to be filtered before being coupled to an antenna and put on air.

Squelch: A control that can be set to silence a receiver it when no signal is being received. This makes it possible to have a receiver operating without being disturbed by band noise. The penalty is that the receiver becomes less sensitive to weak signals, which may trip the squelch intermittently or not at all. Also *mute*.

Squelch line: A connection in a receiver that changes voltage when a signal has been received. This may be used to activate an LED or relay. Most common in VHF and UHF FM equipment.

SRI: Sorry. (CW abbreviation)

SSB: Single sideband (or more accurately single sideband suppressed carrier). An amplitude modulated speech mode with only a single sideband and suppressed carrier. Offers narrower bandwidth (approximately 3 kHz) and higher efficiency than AM. SSB's absence of a carrier signal also lessens interference to other stations. SSB can be generated through either the filter or phasing method.

SSTV: Slow Scan Television. A mode used to send still pictures over a narrow (voice channel) radio bandwidth. Requires a computer or mobile phone, a program or app and interface cable. There are both digital and analogue versions of slow scan television.

SSTV Cam: A SSTV receiving station whose output is viewable on the web. This can be useful as a means of remotely viewing pictures you and others send.

Stacked: Refers to two or more beam antennas placed over one other to increase gain and directivity.

Stage: A portion of a circuit that does a specific job. Common examples in radio equipment include amplifiers, oscillators, filters and mixers.

Stand by: A request for other stations to wait and not transmit.

Static crashes: A common term for radio interference caused by lightning storms.

Station: An installation of radio equipment capable of receiving and transmitting.

Stator: The non-moving vanes of a variable capacitor. Opposite to rotor.

Step attenuator: A piece of equipment that allows signals to be reduced in strength by fine accurate steps (often 1dB). This permits various measurements and comparisons such as antenna and power amplifier gain.

STL: Small transmitting loop. See magnetic loop.

STN: Station. (CW abbreviation)

Store and forward: Refers to a device that receives and then stores a voice message or data for a short time before repeating or relaying it. See parrot repeater.

Straight key: An up and down Morse key where characters are formed manually. As opposed to a side-to-side bug or paddle.

Stray capacitance: Unwanted capacitance in a circuit, as may be caused by components or their leads being too close to one another. Stray capacitance can reduce the performance of radio equipment and potentially cause feedback. Cures include moving parts further apart, keeping connections short, re-orienting components or electrically shielding sensitive stages in a circuit from one another.

Stray inductance: Unwanted inductance in a circuit, such as caused by excessive component or connecting cable lead length. This can reduce the performance of RF equipment, or, in extreme cases, stop it working at all. Stray inductance is minimised by keeping leads as short as possible, repositioning nearby inductors to minimise coupling and having proper shielding between sensitive parts of a circuit.

Strength: Part of the RST scale when giving signal reports. A S1 signal is barely perceptible, S3 is weak, S4 is fair, S6 is good, S8 is strong and S9 is extremely strong. S-meters on equipment also often have readings over S9, including 20, 40 and 60 dB over.

Stub: A resonant length of ribbon feedline, coaxial cable, waveguide or other transmission line. Often used to provide RF filtering (especially at UHF and microwave frequencies) or impedance transformation (for HF and VHF antennas). May be open or closed circuited at the end, depending on application.

Sub audible tone: See *subtone*.

Subcarrier: A low-level carrier signal often transmitted as part of a more powerful FM or analogue television transmission used to impress additional information, e.g. sound or slow speed data. Both signals need to be demodulated separately in the receiver.

Sub-band: A band within a band used for a specific mode or activity. Often defined in band plans. CW, SSB, FM and various digital modes typically have their own sub-bands in the band plan. In most countries these are voluntarily mandated *gentlemen's agreements*.

Subharmonic: Refers to a spurious signal or oscillation that is a submultiple of the main or fundamental signal.

Sub-receiver: A second receiver in a high-end transceiver. Allows reception of another signal while your main receiver is tuned elsewhere. This could be handy for hearing both sides of a contact when monitoring split frequency DX activity.

Subtone: Sub audible tone. A low-pitched below voice frequency tone (around 100 - 200 Hz) used for control functions such as to open a VHF/UHF repeater or activate a link. Almost all modern VHF/UHF FM transceivers have this feature. *CTCSS* or *PL* are other abbreviations used to describe this.

Suffix: The last part of your callsign. Normally one, two, three or (occasionally) four letters after the prefix. The suffix could be random letters or something personally significant such as initials. The latter may be possible in countries that allow callsign choice. See *vanity callsign*.

Sunspot cycle: The cycle where the number of sunspots rapidly rises then gradually decays over approximately an 11-year period. High sunspots are associated with high solar activity that ionises the ionosphere, allowing higher frequency radio signals to return to earth. Solar activity is measured with either the sunspot number or 10.7cm solar flux.

Superhet: Superheterodyne. Supersonic-heterodyne. Refers to a receiver that converts incoming signals to an intermediate frequency at which it is easier to filter, amplify and detect. Superhets replaced regenerative receivers to become the most common receiver type from the 1930s.

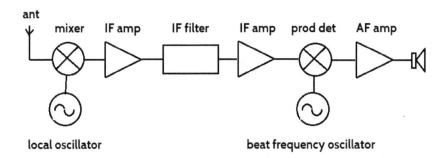

Superregen: Superregenerative receiver. A simple sensitive but unselective type of VHF receiver used by pioneer experimenters in the 1930s, portable operators in the 1960s and short-range remote-control devices today. Receives AM, wideband FM and data signals only.

Surface mount construction: A form of construction on which components are soldered directly to the copper side of a printed circuit board without holes being needed. Special very small 'chip' components are often used. Now dominant in factory made consumer and professional electronics. Opposite to *through-hole construction.*

Surge protector: Device connected in the power lead of electronic equipment to protect against transient power surges.

Sweep generator: A piece of electronic test equipment. Essentially a signal generator that automatically varies (sweeps) its output frequency across a defined range. Useful for testing filter circuits for frequency response.

SWG: Standard Wire Gauge. A pre-metric method of measuring wire thickness. The higher the number the thinner the wire. Also see *AWG.*

Swiss quad: A quad-like beam antenna developed by HB9CV comprising two driven elements.

Switch-mode: A modern type of power supply that uses a *switching regulator.*

Switching regulator: An efficient form of voltage regulator that operates at a high frequency to allow smaller and lighter transformers. Needs to be designed well to avoid causing interference to sensitive audio and radio equipment. Found in switch-mode power supplies.

Switching transistor: A type of transistor designed or used for switching purposes. Some types can also amplify radio frequencies so are used in transmitters.

SWL: Shortwave listener. Common term to refer to people listening to HF radio communication, including broadcasters and amateurs.

SWR: Standing Wave Ratio. Abbreviation of Voltage Standing Wave Ratio. See *VSWR.*

Synchronous detector: A specialised type of detector circuit used in advanced short-wave receivers to provide high quality long-distance AM reception with less fading than otherwise by preserving phase relationships originally in the transmitted signal.

Synthesised: Refers to a receiver, transmitter or transceiver that has a synthesised VFO using a PLL or DDS (as opposed to a VXO or free-running VFO).

Synthesiser: See *frequency synthesiser*.

System Fusion: A mode of digital voice and data communication developed by Yaesu using C4FM 4 level FSK technology. http://www.systemfusion.yaesu.com

T

T-connector: A three-way connector used for power, audio or antenna purposes. Each connector is in parallel with the other two.

T-hunting: Transmitter hunting. See *foxhunting*.

T-match: A type of antenna coupler comprising two variable capacitors and a variable inductor one side of which is grounded. So called as the arrangement of parts resembles the letter T.

Tagstrip: A piece of plastic with several metal solder tags attached as a means of making electrical connections or mounting components. Common in the vacuum tube era, tagstrip construction is useful for simple projects using leaded components where a printed circuit board is not justified.

Tail: The brief signal emitted by a repeater just after a station has finished transmitting. May or may not include a beep.

Tail-ending: An approach to getting contacts where you wait for a contact to finish and then call one of the stations. Most likely they will still be listening and reply. A highly effective method especially if using modest antennas and/or low power.

Tank circuit: A resonant tuned circuit, comprising an inductor and capacitor, used to provide selectivity for frequency-critical applications such as a receiver's front-end, RF oscillator or power amplifier in a transmitter.

Tantalum capacitor: A compact polarised capacitor used in applications where medium to high values of capacitance are required such as audio equipment. Prone to explode if subjected to above the rated voltage. Typical values are 100 nF to 100 uF.

Tap: Tapping point. A point, spike or socket on an inductor that accepts a lead or clip so that inductance can be varied. Common on home wound inductors such as for crystal sets and antenna couplers.

Tapered reamer: A useful hand tool used for enlarging holes in chassis or panels, such as may be required for a control, socket or vacuum tube. A nibbling tool and file can finish the job off if a square or rectangular opening is needed.

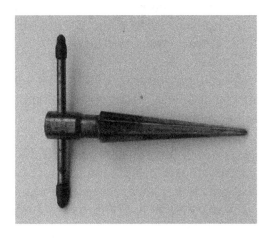

TCXO: Temperature Compensated Crystal Oscillator. An oscillator module often provided in an amateur transceiver to keep its frequency stable even if the external temperature varies. While formerly provided as an optional extra, TCXOs are now essential for modern digital modes that demand high frequency stability.

Technician License: The entry grade of amateur licence in the USA. Allows full VHF/UHF privileges but only limited HF access.

Telegraphy: Relates to Morse code (or telegraph) communications.

Telemetry: Relates to the automated gathering and sending of measurement data via radio, especially from a remote location such as a space probe.

Telephony: Voice communication. As opposed to data or telegraphy.

Temperature coefficient: Refers to the extent to which an electronic component changes value with temperature. A negative temperature coefficient means its value drops with rising temperature while a positive temperature coefficient means the value increases with temperature. Occasionally this is useful but in most cases a component that does not change value with temperature (i.e. zero temperature coefficient) is preferred.

TEP: Trans-equatorial propagation. Refers to a mode of radio signal propagation that allows VHF signals to be propagated long distances across the geomagnetic equator.

Terminating resistor: A resistor connected to the end of a wire antenna such as a rhombic to make it radiate in one direction only. In other applications terminating resistors are used to prevent reflections on a transmission line and as 'dummy loads' when testing transmitter or amplifier stages.

Test point: A point on a circuit board designed to permit easy connection of a multimeter or oscilloscope probe to allow voltages to be measured or waveforms to be inspected for testing or adjustment purposes.

Test pool: A bank of questions, often publicly available, that will be drawn from to produce an amateur exam. (USA)

Test transmission: A transmission for the purpose of testing or adjusting equipment or antennas. You would identify by announcing "<your callsign> testing".

TFT: Thin film transistor. A type of panel often used on flat liquid crystal display screens.

Thermal runaway: A destructive characteristic where parts such as transistors can heat up, cause even more current to be drawn and overheat until destroyed.

Thin: Refers to transmit audio that lacks bass or 'body' so is hard to understand. Can be due to a bad microphone or one unsuitable for the speech amplifier stages in a transceiver.

Third order intercept point: A performance measure pertaining to the ability of equipment such as receivers to handle strong signals without overloading or distorting.

Third party traffic: Messages on behalf of others handled by radio amateurs as a community service.

Three-terminal regulator: A simple type of voltage regulator with three pins, typically labelled input, output and control (or earth). Common examples include the 7805 (+5 volts), 7812 (+12 volts) and LM317 (variable).

Through-hole construction: A method of construction with etched printed circuit boards with holes for component leads to pass through. Largely replaced by surface mount construction which is smaller and cheaper.

Thumbnail switch: A type of small multi-position side-mounted rotary switch designed to be controlled by the user's thumb. Commonly including a 0 to 9 indicator, banks of three or four were used to select and show the frequency of 1970s and 1980s VHF equipment. Still used in test and professional equipment.

Ticket: Popular term for an amateur certificate or licence.

Tilt over: Describes an antenna mast or tower whose top can be tilted over, allowing work on antennas to be more safely done near ground level.

Time constant: Refers to the charge time of a resistor-capacitor combination for the voltage across the capacitor to reach 63.2% of the voltage applied as measured in seconds. The larger the resistor and capacitor values the longer the time constant. For a non-electrical analogy think of a high value resistor as a thin pipe and a high value capacitor as a large water tank, with the time constant being the time to mostly fill the tank.

Time out: A timer on a repeater that causes it to shut down if a user's transmissions exceed a certain duration (typically 3 to 5 minutes). Intended to prevent monopolisation of the repeater. Also see *TOT*.

Tin: To apply a thin coating of solder to a component lead or connection to allow easier soldering.

TKS: Thanks. (CW abbreviation)

TNC: Terminal Node Controller. An interface between a computer terminal and radio transceiver used for transmitting and receiving packet radio.

TNX: Thanks. (CW abbreviation)

Toggle switch: A (typically) two position switch using a lever. Commonly used to switch power and other functions in older equipment.

Tolerance: The extent to which it is acceptable for a component's measured resistance, capacitance or inductance to vary from that shown on the case. Typical components have 5, 10 or 20% tolerances with precision components being 1%.

Tone burst: A short whistle or burst of tone, typically at 1750 Hz, transmitted to activate an amateur radio repeater or link. Mostly used in Europe. Other countries such as in North America and Australasia use less obtrusive subtone systems such as CTCSS or PL instead.

Top band: The amateur 160 metre (1.8 MHz) band. (UK)

Top loaded: Refers to a vertical antenna with a loading coil near the top of the antenna. This is mechanically inferior but delivers better performance than base loading.

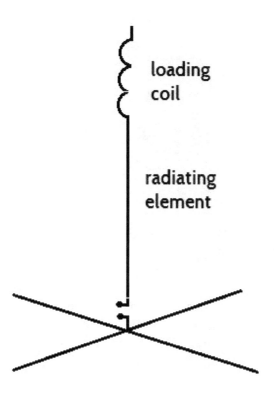

loading coil

radiating element

Toroid: A doughnut-shaped ring of ferrite or iron powder that wire can be wound on to form an inductor or transformer. Toroids save space compared to cylindrical air-wound coils due to their permeable and self-shielding properties. They are commonly used in tuned circuits, antenna baluns and power transformers.

Toroidal: Refers to an inductor, balun or transformer comprising wire or wires wound on a toroid (as opposed to a solenoid or cylindrical former).

TOT: Time Out Timer. A feature found in many VHF / UHF transceivers that shuts off transmission if the user has been talking longer than a set time. The user normally receives a warning sound before the time out operates. This function prevents the transmitter from overheating and gives the user warning if using a repeater that may also have a TOT or 'alligator'.

Tower: A support structure for a radio antenna. May support wires or rotatable beams. Especially on 80 and 40 metres some amateurs feed the tower itself to form a vertical antenna element.

T/R: Transmit/Receive, especially when referring to a transmit/receive switch.

Track: Conductive copper trace on a printed circuit board. Components are soldered to circuit board tracks to form connections between them.

Traffic: Messages carried by amateur radio, often as a public service. (USA)

Transceiver: A transmitter and receiver in the one package to allow sending and receiving of radio signals. Sharing of components reduces production costs and makes operating easier compared to separate transmitters and receivers.

Transducer: An electrical component that converts movement to electrical energy or vice versa. Dynamic microphones and speakers are examples.

Transformer: A component comprised of windings of wire that allows voltage or impedance step up or step down such as required in power supply, amplifier and antenna circuits. See *balun*.

Transient: A momentary sudden flow of energy such as may appear on a power rail when there is a rapid change to an item's current consumption (e.g. it being switched on and off). Transients can be filtered to protect sensitive circuits and eliminate radio interference caused. Also known as voltage, current or energy spikes.

Transistor: A solid state three terminal amplifying and oscillating device. A small current on the base allows control of a larger collector – emitter current, thus providing a switching or amplifying effect. The ability to amplify also allows for positive feedback which can make the transistor oscillate at a frequency controlled by a tuned circuit or crystal. Transistors largely replaced electron tubes in the 1960s due to their small size, robustness and simpler power supply requirements.

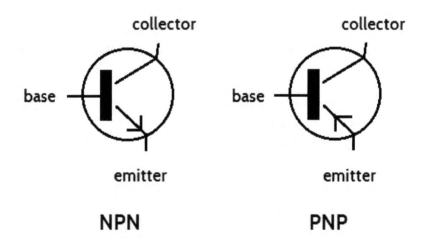

Transmatch: An accessory containing variable capacitors and inductors, connected between a transceiver and the antenna, that allows impedance to be transformed such as may be required for efficient transmitting and receiving. Also *antenna coupler*.

Transmission line: A line that carries radio signals, especially the feedline between the antenna and transceiver. Common types include coaxial cable, ladder and open wire. More broadly, the term applies to any line that carries electricity or signals over long distances, such as for power grid and telephone networks.

Transmitter: A generator of radio frequency energy used to broadcast a signal.

Transverter: An accessory that allows a transceiver to operate on a frequency it is not designed for by using a frequency converter to add or subtract frequencies. Often used to allow operation on LF or microwave bands with commonly available HF or VHF transceivers.

Trap: A tuned circuit that allows signals of particular frequencies to be rejected or isolated. Typical applications include use in antennas to allow operation on multiple bands or receiver front-end filters to reduce breakthrough from strong out of band signals.

Trap dipole: A dipole antenna containing traps part way along the elements to allow operation on two or more bands.

TRF: Tuned Radio Frequency. Refers to a type of 'straight' AM radio receiver comprising a radio frequency amplifier, detector and audio amplifier popular in the early days of radio. Sometimes the detector may be made regenerative to increase gain or permit CW reception. Unlike a superhet a TRF receiver has no intermediate frequency and was difficult to make selective especially if a wide tuning range was required.

Triac: A type of semiconductor that (unlike an SCR) will pass current in either direction when triggered by current applied to its gate terminal. This makes them useful in electrical switching and control circuits such as light dimmers and motor speed controllers.

Triband: Refers to equipment or antennas (especially) that can operate on three frequency bands.

Trickle charger: A type of battery charger that feeds only a small current into a rechargeable battery. Charging time may be overnight or longer. Improved battery technology has sped charging times so trickle chargers are less common than previously.

Trifilar transformer: Type of transformer with three wires twisted together before the combination is wound on a rod or toroid. Common in broadband RF circuits such as transmitter power amplifiers and 1:1 antenna baluns.

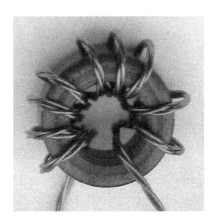

Trimcap: Trimmer capacitor. A variable capacitor that can be adjusted with a screwdriver or pliers. Once correctly adjusted it can be left alone. Used to align critical circuitry inside transmitters and receivers. See *Preset*.

Trimpot: Trimmer potentiometer. A variable resistor that can be adjusted with a screwdriver. Once correctly adjusted it can be left alone. Commonly used to adjust levels, thresholds and bias voltages in electronic equipment. See *Preset*.

Triple conversion: Relates to a superhet receiver in which incoming signals have three radio frequency conversions (that is three intermediate frequencies) before being converted to audio.

Tropo: Tropospheric ducting. Conditions encountered on the VHF/UHF bands, often due to a temperature inversion, that allow extended range communication over hundreds or even thousands of kilometres. Can be forecast by looking at weather maps. See *Hepburn Index*.

Troposphere: A layer in the atmosphere that can facilitate extended range VHF and UHF radio communications under certain conditions such as during temperature inversions. Inversions can cause ducts to be form, carrying VHF radio signals hundreds of kilometres beyond their usual 'line of sight' range.

TS: Letters that precede the model number of Kenwood brand HF transceivers.

TTL: Transistor-transistor logic. Older style of digital logic IC.

TU: Thank you. (CW abbreviation)

Tuneable IF: An IF and audio stage in a receiver that is tuneable. A crystal-controlled converter can then be added to enable it to receive a desired band. This arrangement was common in the days when it was difficult to obtain high IF selectivity and/or frequency stability at higher HF and VHF frequencies.

Tune for maximum smoke: A funny saying, popular amongst 'screwdriver experts', indicating that one should adjust all settings in a piece of equipment for maximum audio or RF power output regardless of the cleanliness or quality of the output.

Tuned circuit: A circuit comprising an inductor (or coil) and capacitor that allows a particular frequency to be passed or rejected. Tuned circuits have their parts in series or parallel, depending on the application.

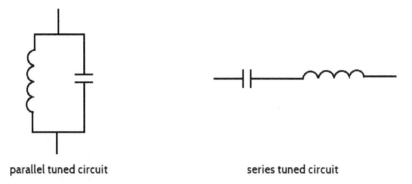

parallel tuned circuit   series tuned circuit

Tuning range: The range of frequencies that a receiver can cover. Same as frequency coverage.

Tuning rate: The number of kilohertz per turn of a radio's tuning knob. A fast tuning rate makes signals hard to tune in while a slow tuning rate makes tuning tedious. This can be adjusted on modern transceivers.

Tuning step: The spacing between adjacent frequencies when you turn the tuning knob of a transceiver. Fine tuning steps are desirable if tuning in narrowband signals while coarse tuning steps are adequate if tuning in a wide signal (e.g. AM).

Tuning up: Refers to adjustments to an antenna coupler etc to ensure a good impedance match between the transmitter and the antenna system. This is required for efficient power transfer to non-resonant antennas.

Tunnel diode: A rare type of diode whose 'negative resistance' characteristic allows them to operate as oscillators and amplifiers. Tunnel diodes can function at microwave frequencies due to their low capacitance.

Turns: Individual windings of wire on a solenoidal coil, toroidal inductor or transformer.

Turns ratio: The ratio between the number of wire turns on one winding and the number of wire turns on another in an inductor, transformer or balun. The turns ratio determines a transformer's voltage step-up or step-down ratio. Impedance transformation is similar except the ratio is equal to the ratio between the squares of the number of turns on each winding.

Turnstile: A type of VHF or UHF antenna comprising two half wavelength dipoles connected in parallel. One dipole has a slightly longer feedline than the other, causing the signal from the transmitter to arrive at it slightly later than for the other dipole. This timing (or phase) difference causes a circularly polarised signal to be radiated. This is desirable for communication via satellites (whose antennas are spinning rapidly through space so do not exhibit constant horizontal or vertical polarisation).

TVI: Television interference. Interference caused to a television set due to the presence of a radio transmitter. This may be caused by spurious emissions from the transmitter or because the TV set has insufficient immunity to out of band signals.

Tweak: The art of aligning the internal circuits of an electronic item, such as a transmitter or receiver, for maximum performance. This is both a legitimate activity and one partaken by 'screwdriver experts'.

Tx: Transmitter or transmitting. Opposite to Rx.

Tx Factor: UK-based amateur radio video program.

Txcvr: Transceiver.

Type approved: Refers to transmitting equipment that has been approved by your country's spectrum regulator. Because they have passed a technical exam, radio amateurs are one of the few spectrum users permitted to operate equipment that has not been formally type approved.

U

uBitx: A popular kit-type multiband band SSB transceiver developed by Ashhar Farhan VU2ESE. www.hfsignals.com/

UHF: Ultra high frequency: 300 – 3000 MHz. Includes the amateur 70cm, 33cm and 23cm bands. Ideal for local and satellite communications. Extended distances are possible with changed atmospheric conditions, e.g. a temperature inversion.

ULS: Universal Licensing System. Online database used by FCC to manage amateur and other radiocommunication licensing. (USA)

Unbalanced feedline: A form of feedline that is unbalanced. Coaxial cable is the most well-known example. This is robust and can be run alongside metal objects such as antenna masts and gutters. It is easy to use in that 50 ohm coaxial cable can conveniently plug in to a transceiver whose output is also unbalanced. However, it can be lossy where there are large impedance mismatches or if there are long runs, especially on VHF and UHF frequencies.

Unchannelised: Refers to amateur bands and/or modes whose users operate on any frequency within a band or sub-band limit rather than confine activity to specific channels. HF and VHF CW and SSB are examples of unchannelised activity.

Under-deviating: Refers to an FM voice transmission whose excursion from its centre frequency is insufficient for the characteristics of the receiver detector it is used with. This causes the signal to sound quiet and be less readable than fully deviated transmissions. Talking louder into the microphone helps slightly but the real cure is an internal adjustment or modification to increase deviation.

Under-driving: Applying insufficient power required to fully drive an amplifier stage. The result is likely to be reduced audio or RF power output. Unlike over-driving, under-driving is rarely noticed unless severe.

Under-modulation: Applying insufficient audio to a transmitter modulator stage. The result is likely to be a weak signal and/or faint audio. Unlike over-modulation, under-modulation is rarely noticed unless severe. However, it should be fixed if you want your signal as readable as it can be for the output power used.

Under-rated: A component with a voltage or power rating less than required for a circuit. For instance, a 12-volt power supply fitted with a 10-volt rated filter capacitor. The under-rated component will be placed under stress and will not last long. It could even explode. A proper substitute would be rated significantly higher – e.g. 25 or 30 volt.

Unity gain: Refers to an antenna that has no gain relative to a reference antenna. This is usually an isotropic radiator or half wavelength dipole depending on the reference point.

Unun: Unbalanced to unbalanced. An impedance transformer often used between (unbalanced) coaxial feedlines and unbalanced antennas, especially end and off-centre fed wires.

Up [number]: Short for 'listening up [number] kHz'. Means that the calling station will be listening for calls [number] kHz above their transmitting frequency as part of split frequency operating. You need to adjust your transmitter to call them there, while leaving your receiver untouched. This technique is commonly used by sought-after DXpedition stations to manage large pile-ups. It is common for CW DX-peditions to listen 'up 1 kHz' and SSB expeditions to listen 'up 5 kHz'. See *split frequency*.

Up conversion: A mixing scheme in a receiver or transceiver where incoming signals are converted to a high first IF, normally in the low VHF range. This is often then converted to a lower second IF for filtering and detection. This arrangement is particularly popular in general coverage HF equipment where an IF that is within the receiver's tuning range would cause problems.

Upconverter: A radio circuit comprising a local oscillator, mixer and filters that converts a lower frequency radio signal to a higher frequency signal. Commonly used to allow MF and HF reception on SDR dongles that by themselves may only cover VHF and UHF frequencies. Opposite to downconverter.

Up-down buttons: Buttons, often provided on a microphone or transceiver, to allow a setting such as frequency, channel or volume to be changed.

Upgrade: The act of doing a more complex exam to obtain a more advanced category of amateur licence and obtain more operating privileges.

Uplink: A transmission from ground to a satellite.

Uplink frequency: The frequency you transmit on to communicate via a satellite.

UR: You are. Your. (CW abbreviation)

USB (computer connection): Universal serial bus. A common type of socket on a computer used to connect accessories including memory sticks and SDR dongles.

USB (transmitting mode): Upper sideband. An SSB transmission where the audio component extends above the frequency of the suppressed carrier. Normally used on amateur bands above 10 MHz. Opposite to LSB.

UTC: Coordinated Universal Time. Also known as Greenwich Mean Time. International time used by amateurs to avoid confusion arising from local time zones or daylight-saving time.

UV: Ultra-violet. Refers to the portion of the electromagnetic spectrum with a frequency higher than that of the visible light spectrum but lower than x-rays.

V

Vacuum tube: Electron tube. Valve. The first electronic component capable of amplification and oscillation, comprising a heater, cathode, anode and at least one grid in an evacuated glass envelope. Normally requires at least two power sources to operate – low tension for the heater and high tension for the anode. Largely superseded by transistors but retain a following amongst some audio and RF amplifier enthusiasts.

Vacuum variable capacitor: An expensive and specialised variable capacitor able to withstand high voltages such as required on magnetic loop antennas and high power RF amplifiers.

Valve: Alternative name for vacuum tube (UK, Australia, NZ).

Vanes: The parallel metal blades of a variable capacitor. There are two sets – fixed and moving, with the former connected to the shaft and often the frame of the variable capacitor.

Vanity callsign: Refers to a callsign that has not been randomly allocated to you. That is, you choose letters that are personally significant, e.g. your initials. You may have to pay a special fee for this service. (USA)

Varactor diode: A diode designed to vary in capacitance when voltage applied to it is varied. The variable voltage may be supplied through a manual control such as a potentiometer or automatically for example in a phase locked loop circuit. Varactor diodes are often used in simple receivers and FM transmitters. Can be used with a crystal or ceramic resonator to adjust a transmitter or receiver over a limited tuning range with a potentiometer. It is worth noting that some non-varactor types of diode also have a variable capacitance effect, for example power diodes.

Variable capacitor: A capacitor that can be varied in value by rotating a shaft that is connected to moving plates that are electrically insulated from a set of fixed plates. The more the overlap the greater the capacitance. Variable capacitors usually have minimum values around 5 or 10 pF and maximum values between about 50 and 500 pF. Commonly used in receiver tuning controls and antenna couplers. The example pictured below is single gang (or section) although two and three gang capacitors also exist.

Variable resistor: A resistor that can be varied in value by adjusting a knob. Commonly used in audio volume and level controls. It may have either three connections (potentiometer as pictured below) or two connections (rheostat).

Varicap diode: See *varactor diode*.

Variometer: A type of inductor comprising a small coil rotatable inside a large coil. Depending on how these are connected, this arrangement can form a variable inductor or allow coupling between stages in a transmitter or antenna tuner to be varied. Variometers date from the early years of radio and are most used in low and medium frequency equipment.

VCO: Voltage controlled oscillator. An oscillator whose frequency can be varied by adjusting a voltage. A common application is use in a phase locked loop VFO.

VE: Volunteer examiner. A person who administers amateur licence examinations. (USA)

VEC: Volunteer examinations coordinator. An organisation approved by the Federal Communications Commission to run the administration of amateur licence examinations through *VEs*. (USA)

Vee beam: A type of HF wire antenna comprising two wires at an acute angle to one another shaped like a letter V when viewed from the air. The antenna is fed with open wire line where the two wires almost meet. another fed where they meet. Vee beams can provide substantial gain in two directions. Their main disadvantages compared to the more commonly used yagi include the large land area required and the inability to rotate them without adding wires and elaborate switching.

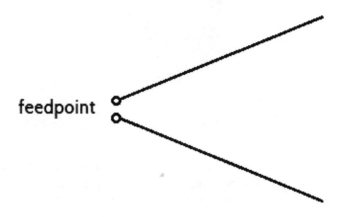

feedpoint

Velocity factor: Relates to a property of antenna feedlines for signals up them to travel at a slower speed than light. Coaxial cable typically has a velocity factor of about 0.66 while open wire feedline may have a velocity factor of 0.98.

Vernier dial: A radio dial that contains gears to offer mechanical reduction and make adjustment of variable capacitors as used in tuning controls easier. Widely used in receivers and transceivers before DDS VFOs took over.

Vero board: A 1970s type of matrix circuit board that has copper strips connecting rows of holes. This can make the construction of some projects easier. Strips can be broken with an old drill bit if electrical isolation is required. Vero board is most suited to DC and audio projects as capacitance between the tracks can impair operation at HF and VHF radio frequencies.

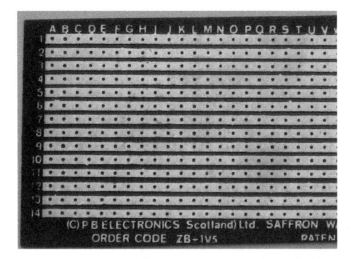

Vertical: A common name for a single element omnidirectional vertically polarised antenna, for example ground planes and J-poles.

Vertically polarised: Refers to an antenna that emits or responds best to vertically polarised signals, i.e. those whose lines of electric flux are in the vertical plane.

VFO: Variable Frequency Oscillator. Refers to the main tuning control on a receiver, transmitter or transceiver that allows the operating frequency to be changed, i.e. a tuning control. A generic term that applies to both free-running and synthesised oscillators.

VHF: Very high frequency. 30 – 300 MHz. Portion of the radio spectrum suitable for local and satellite communication. Includes the amateur 6m and 2m bands. Normal range is up to about 100 kilometres with more efficient SSB and digital modes giving several times that. Extended distances are also possible with changed atmospheric conditions, e.g. a temperature inversion, or, in the lower VHF range, sporadic-E propagation.

Vibrator: A now obsolete electromechanical technique of generating high voltages from low DC voltages as formerly required to operate tube equipment from a vehicle battery. Common in pre-transistor vehicle-mounted communications equipment. The modern approach would be to use a switching regulator.

Vintage: Refers to old equipment, e.g. boatanchors. Definitions are somewhat loose and some eBay sellers even refer to 1980s and 1990s items as 'vintage'.

VK2ABQ: A type of compact square-shaped multiband HF beam antenna comprising two wire elements bent in on themselves. Related to the Moxon. Named after its inventor Fred Caton VK2ABQ who first described it in "Electronics Australia" magazine.

VLF: Very low frequency. Frequencies between 3 kHz and 30 kHz.

Voice keyer: A voice recorder/player connected to a transmitter and timer. Allows you to call CQ automatically and respond, when stations reply, with your own voice. Often used in contests.

Vol: Volume control. Controls audio output from receiver or audio amplifier.

Volt: Unit of electromotive force. 1 volt is the amount required to cause a current of 1 ampere to flow through 1 ohm of resistance.

Voltage balun: A type of balun that forces equal (but opposite in phase) voltages to each of its balanced side connections (assuming a signal is being applied to its unbalanced side).

Voltage divider: A network of two series connected resistors that divides the voltage present at their junction according to the radio of their values. This lower voltage is often useful in transistor and IC circuits. A potentiometer can form a variable ratio voltage divider such as useful for level and volume controls.

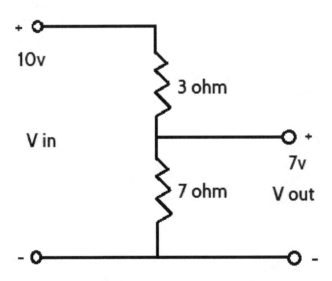

Voltage drop: Refers to an undesired reduction in voltage supplied to equipment due to a power supply or cable being too small, thin or long for the job. Voltage drop is most apparent when high current is being drawn, such as an SSB transmitter on voice peaks. Minor voltage drops (e.g. 0.5 volt) are acceptable but much more than that can cause distortion on transmit (FMing on SSB or chirp on CW). Treat voltage drop by using short thick power leads, substituting a fully charged battery and/or reducing transmit output power.

Voltage multiplier: Circuitry comprising diodes and capacitors to multiply the voltage applied. Often used to obtain a higher voltage from a lower voltage transformer winding.

Voltage reference: A precision power supply that provides a known and stable voltage as may be required for testing and measurement purposes.

Voltage regulator: Component or circuitry in a power supply that maintains a constant voltage despite the current drawn by the load varying. The simplest type is a Zener diode. More advanced are linear voltage regulator ICs such as the 7805 pictured below useful for many IC circuits. These are simple to use but can be inefficient, particularly as the input voltage rises. More efficient (but also more complex) are lightweight switching regulators as used in modern computer and consumer equipment.

Voltage spike: A short burst of high voltage on a power connection. Has potential to harm equipment. Components such as MOVs can be used to offer protection.

Voltmeter: Instrument that measures voltage. Voltmeters were once common stand-alone instruments but today are most commonly incorporated in multimeters.

VOM: Volt ohm meter. See *multimeter*.

Voting receiver: A part of an advanced repeater system with receivers at multiple locations. The control unit of the repeater 'votes' which receiver is getting the strongest signal from the transmitting station and feeds the winning receiver's output to the repeater's transmitter. This improves the quality of signals heard through it, especially from handheld or mobile stations.

VOX: Voice operated transmit. An audio sensor circuit that turns on the transmitter when you start speaking into the microphone. A common feature on most HF transceivers that enables more spontaneous transmissions without you having to press the *PTT*.

VSWR: Voltage Standing Wave Radio. An indication of the amount of impedance mismatch between an antenna and the feedline connected to it. A 1:1 VSWR indicates a perfect match, 2:1 indicates a mismatch where impedances are out by a factor of two (i.e. either half or double), 3:1 out by a factor of three and so on. VSWR is a somewhat crude measurement since it does not indicate the direction impedances are out by (i.e. whether the antenna is a lower or higher impedance than the feedline). Neither does it indicate antenna efficiency or adequately express reactive loads. Nevertheless, VSWR is handy for simple tests, such as determining whether anything has changed on a known good antenna or when cutting an antenna to length.

VSWR meter: An instrument used to determine that an antenna is correctly cut for a desired operating frequency or an antenna coupler is properly adjusted. Either built in to modern transceivers or available as an external unit. The meter pictured below is a 1970s deluxe model popular with 27 MHz CBers but also usable on the amateur bands.

VTVM: Vacuum tube volt meter. A test instrument using vacuum tubes and a meter movement to measure voltage and often also current and resistance. A multimeter would now be used for such measurements.

VXO: Variable crystal oscillator. A crystal oscillator circuit whose frequency can be varied slightly by adding a variable capacitor and/or inductor in series with the crystal. Often used in simple transmitter and receiver circuits to provide limited frequency agility.

VY: Very. (CW abbreviation)

VY 73: Very best wishes. (CW abbreviation)

W

W: Watt. Unit of electrical power as used to indicate power usage or output from an amplifier or transmitter. Equals volts x amperes.

WARC: World Administrative Radio Conference. Former name of ITU conference to agree on radio spectrum allocation and administration including for amateur radio. Now known as *World Radio Conference*.

WARC bands: Refers to new HF bands amateurs were granted after the 1979 World Administrative Radio Conference (WARC). These are at 10, 18 and 24 MHz. By international agreement WARC bands are contest-free.

WAS: Worked All States. A popular award certificate for confirming contacts with stations in all 50 US states.

Waterfall: A type of display view that allows you to view activity over time on a slice of radio spectrum on a computer screen. Signals appear as bright or coloured lines. Such displays are common on software defined receivers and with digital mode software such as *Digipan*.

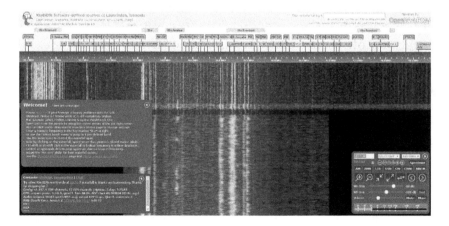

Wattmeter: Instrument that measures power. Most commonly RF output from a transmitter. Often combined with a VSWR meter.

Waveform: A visual depiction of how an AC electrical signal varies with time such as observed on an oscilloscope.

Waveguide: A hollow rectangular or circular metal pipe used to convey microwave radio signals with minimal loss.

Wavelength: The length of a wave. Those at radio frequencies are measured in metres, centimetres or millimetres. Lower frequencies have longer wavelengths and higher frequencies have shorter wavelengths. Often represented by the Greek lambda (λ) symbol. A quick formula to work out wavelength (in metres) from frequency is wavelength = 300 / frequency (in MHz). Amateur bands are often referred to in wavelength terms, e.g. 160, 80, 40 metres etc for 1.8, 3.5, 7 MHz etc. This makes the estimation of antenna lengths easier. For example, a half wavelength dipole antenna for 80 metres (3.5 MHz) is approximately 40 metres end to end while a quarter wavelength ground plane for 2 metres (144 MHz) has elements about 50cm long.

WAZ: Worked All Zones. A popular DX award issued by CQ magazine to amateurs who have contacted stations in all 40 CQ zones.

WBFM: Wide band frequency modulation. Refers to an FM transmission with a wide deviation and bandwidth such as 88 – 108 MHz broadcast stations and (previously) analogue TV sound. Rarely used by amateurs due to its bandwidth. Opposite of *NBFM*.

Weak signal modes: Term used by VHF and UHF operators to refer to modes that are efficient even when signals are weak. For instance, CW, SSB and many digital modes. In contrast FM voice is not a weak signal mode. For a given transmitter output power weak signal modes can be detected over longer distances than other modes.

Web receiver: Radio receiver controllable via the internet. Open access types can be accessed by anyone who can hear how signals (including their own) sound like from a different location. A larger number can be accessed from http://www.sdr.hu Also *web SDR*.

Web SDR: Software defined radio controllable via the internet. See *Web receiver*.

WeFax: Weather fax. An analogue mode for the transmission of meteorological reports and weather charts over radio.

Wet cells: Refers to the contents of rechargeable batteries with liquid electrolyte such as lead acid types used in vehicles.

White noise: Refers to a constant noise with equal amplitude across a wide range of frequencies. Useful for testing audio circuits, white noise can be produced by either a dedicated signal generator or an FM broadcast receiver tuned between stations.

WID: With. (CW abbreviation)

Wide: Refers to a signal that has a bandwidth excessive for its mode. For example, an SSB signal that is 10 kHz wide when it should be about 3 kHz. Caused by faulty equipment or adjustment.

Wideband: Of or having a wide bandwidth. Examples include high-speed data transmissions or an antenna that can be used over a wide frequency range. Also broadband. Opposite to *narrowband*.

Winlink: A worldwide messaging system that allows email and data files to be sent via amateur radio.

Wire wound resistor: A type of resistor, comprising a coil of resistance wire, used for high power DC or low frequency AC applications. Their inductance makes them unsuitable for use in RF circuits. Often in a rectangular or cylindrical housing, their value is normally printed on case instead of the colour code used to identify most other types of resistor.

Wire wrap: A largely obsolete form of solderless connection based on tightly wrapping special wire around pins with a special wire wrap tool. Initially used in telephone exchanges, the technique spread to electronics in the 1960s and 1970s.

Wireless telegraphy: Original name for radio communication, which, in the time the term was used, was mostly by Morse code.

WKD: Worked. Made radio contact with. (CW abbreviation)

WKG: Working. Making contact with. (CW abbreviation)

WL: Will. (CW abbreviation)

Woodpecker: The nickname for a Russian over-the-horizon radar that caused widespread interference to HF communications. So-called due to its sound on the air.

Wouff Hong: A sacred symbol in US amateur radio lore standing for civility and order on the airwaves. The actual device is a fearsome wooden instrument of torture displayed at the ARRL headquarters. See *Rettysnitch*.

Work (someone): To have an amateur radio contact with someone.

Working conditions: A description of your station, e.g. transceiver, antenna and output power.

World Radio Conference: International conference, organised by the ITU, where agreement is reached on radio spectrum regulations and frequency allocations including for amateur radio bands.

WPM: Words per minute as in Morse code sending speed. 5 wpm is slow, 15 wpm is moderate speed, 25 wpm is fast.

WSJT: Weak Signal by JT (Joe Taylor). A computer program enabling weak signal digital communication in conjunction with an SSB transceiver developed by Joe Taylor K1JT. https://physics.princeton.edu/pulsar/k1jt/

WSJT-X: A computer program offering extended features compared to the original WSJT. These include support for multiple weak signal modes including FT8, JT4, JT9, JT65, WSPR and more.

WSPR: Weak signal propagation reporter. A slow speed low bandwidth digital mode most useful for low power beaconing and propagation study. Can be transmitted with an HF SSB transceiver, audio connection, computer and software or with dedicated WSPR equipment. The WSPRnet website allows you to see what radio communication paths are currently open around the world. http://www.wsprnet.org

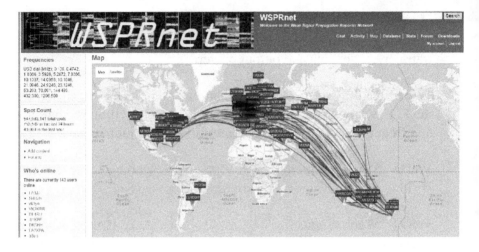

WWFF: World Wide Flora and Fauna. An awards program that promotes portable activity from nature parks and reserves around the world. http://www.wwff.co

WWV/WWVH: Two HF time and frequency standard stations in the United States. Audible all over the world on 2.5, 5, 10, 15 and 20 MHz, they allow accurate time and frequency setting of equipment.

Wx: Weather. Amateurs are often known for talking about the weather so it is only convenient that there is an abbreviation for it, even if its use doesn't save much time on speech modes.

W8JK: A type of wire antenna comprising two closely-spaced dipole elements fed out of phase. Normally bidirectional but it can be made unidirectional with the correct phasing between elements. Typically fed with open wire feedline, W8JKs are ideal for the experimenter who enjoys a challenge.

X

X-beam: A type of directional 2 element beam antenna occasionally built by experimenters. The director and reflector form the shape of an X, allowing a single point of support and eliminating the need for a boom.

XCVR: Old fashioned abbreviation for a transmitter.

XIT: Transmitter incremental tuning. A control that allows you to slightly vary the transmit frequency without changing the receive frequency.

XMTR: Old fashioned abbreviation for a transmitter.

XTAL: Old fashioned abbreviation for a crystal.

XYL: Amateur parlance for wife (ex young lady).

Y

Yaesu: A popular Japanese manufacturer of amateur radio equipment. https://www.yaesu.com/

Yagi: A common type of directional gain beam antenna with tubular or wire elements (reflector, driven, directors) arranged parallel to one another. The more the elements the sharper the beamwidth and the higher the gain.

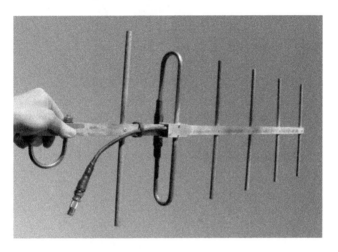

YL: Young Lady. Female ham (any age).

Z

Z (impedance): Impedance. High-Z = High impedance. Low-Z = Low impedance.

Z (time): Zulu time. Refers to UTC or GMT. E.g. 1000z is 1000 UTC.

Z-match: A form of antenna coupler often built at home. Making use of a tapped inductor and a dual gang variable capacitor it can not only transform impedances but also reduce harmonics and protect the receiver from strong out of band signals. Z-matches also uniquely allow operation over a wide range of frequencies without switching the inductor.

Zener diode: A special type of diode used in voltage regulation and over-voltage protection circuitry. Zener diodes start to pass current against the direction of the arrow in the symbol when the voltage applied exceeds its rated breakdown voltage. Once current is flowing the voltage across the diode remains constant, even if that applied rises. Zener diodes are available in various voltages from less than three to over 30 volts and can be wired in series if a higher voltage is desired.

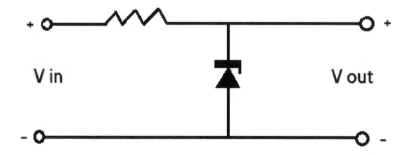

Zepp: A old type of wire antenna commonly fed with open wire feedline, so-called due to their use on air ships (Zeppelins). Can be either end or centre-fed.

Zero beat: Tuning an SSB or CW receiver onto a transmission containing a carrier (e.g. CW or AM) so that no beat note is heard. This is done to check a transmission's carrier frequency, tune a transmitter to a signal's exact frequency or test a VFO's frequency stability.

ZL special: A type of phased array directional beam antenna comprising two driven elements spaced about 1/8 wavelength apart.

Zone: One of a number of ways to geographically divide the world for the purpose of awards or certain contests. The two main zone systems in use are CQ zones and ITU zones.

ZS6BKW: A version of the G5RV HF doublet antenna that is easier to match on some bands.

3 DEFINITIONS: NUMBERS

¼ **wavelength**: A common element length for vertical antennas such as the ground plane as popular on the higher HF, VHF and UHF bands. Such antennas exhibit a low impedance suitable for feeding with coaxial cable.

½ **wavelength**: A common end-to-end length of a half wavelength dipole. Provides a suitable impedance for coaxial cable if fed at the centre. Alternatively, it may be more convenient to feed the ½ wavelength of wire at or near the end. However, its impedance will be high and some form of impedance transformation will be required.

5/8 wavelength: A common length for HF and VHF vertical antennas. While non-resonant it is the length that provides the most radiation closest to the horizon.

1:1 (VSWR): The ideal reading on a VSWR meter. While it indicates an absence of impedance mismatch between transceiver and antenna a 1:1 reading by itself does not necessarily indicate an efficient antenna system as losses can accrue through other means.

1:1 balun: A balun with no impedance transformation properties. Popularly connected between a balanced antenna (such as a half wave dipole) and an unbalanced feedline (e.g. coaxial cable) to prevent feedline radiation.

1.25 metres: The 222 MHz amateur band as available in some North American countries.

2 metres: The 144 MHz band. 144-146 MHz in Europe, 144-148 MHz in North America and Australia. Popular for local communication, this is a band that many beginners have their first contact on. Cheap handheld FM transceivers are widely available for this and the 70cm band.

3 centimetres: The 10 GHz amateur band. Popular with microwave experimenters.

4 metres: The 70 MHz amateur band as available in many European countries.

4nec2: An antenna modelling program. https://www.qsl.net/4nec2/

4 square: An array of four vertical antennas whose directivity can be changed by switching feedline phasing networks in and out. Favoured by 80 and 40 metre DXers.

4:1 balun: A balun transformer with an impedance transformation ratio of 4:1. This is useful in many RF power amplifier circuits or with certain types of antennas.

5 and 9 (or 5 9): Common signal report for a strong signal. Often given in a contest to satisfy the rules requirement for an exchange even if the signal is actually much weaker.

6 centimetres: The 5.8 GHz amateur band. Off-the-shelf low power licence-free data and video transmitters also operate in or near this band.

6 metres: The 50 MHz amateur band. Popular for medium distance communication over summer and long-distance contacts during solar peaks.

9 centimetres: The 3.3 GHz amateur band.

10 metres: The 28 MHz amateur band. Popular for long distance communications during sunspot peaks and medium distance communication over summer.

10.7cm solar flux: A measure of solar activity often used in preference to sunspot numbers. A high flux number (e.g. over 100) ionises the ionosphere and allows it to operate as a reflector of higher frequencies. A low flux number (e.g. under 70) indicates low solar activity and reduced long distance propagation on the higher HF bands.

11 metres: Sometimes used to refer to the 27 MHz CB band.

12 metres: The 24 MHz amateur band (actually just below 25 MHz). Most active around solar peaks.

13 centimetres: The 2.3 GHz amateur band. Microwave ovens operate not far from this allocation.

15 metres: The 21 MHz amateur band. Popular for long distance daytime communication during sunspot peaks.

17 metres: The 18 MHz amateur band. Popular in high sunspot years.

20 metres: The 14 MHz amateur band. Very popular for worldwide communication.

20 over: Short for 20 dB over S9 as indicated on an S-meter. A very strong signal.

23 centimetres: The 1296 MHz amateur band. Popular for amateur television.

30 metres: The 10 MHz amateur band. Popular for worldwide communication using Morse and digital modes.

33 centimetres: The 900 MHz amateur band as available in some countries

40 metres: The 7 MHz amateur band. Popular for daytime medium distance and evening long distance communication.

40 over: Short for 40 dB over S9 as indicated on an S-meter. An extremely strong signal.

50 ohm: The output impedance of amateur transmitters. A good match for 50 ohm coaxial cable such as sold for communications purposes.

60 metres: The 5 MHz amateur band as available in some countries. There are often power, frequency and mode restrictions due to its shared nature.

60/40 solder: Refers to solder that is 60% tin and 40% lead as used for electronic work. Replaced by lead free solder in some countries.

70 centimetres: The 430 MHz amateur band. Popular for short distance communication via repeaters and medium distance communication via SSB, digital modes or using satellites.

73: Expression of best wishes between amateurs. '73s' is often heard but incorrect.

75 ohm: The common characteristic impedance of coaxial cable used for broadcast television.

75 metres: Sometimes used to refer to the top end of the 3.5 MHz amateur band available in some countries. SSB and AM activity may be found here.

80 metres: The 3.5 MHz amateur band, especially the portion below about 3.8 MHz. Popular for short, medium and sometimes long-distance communication at night.

88: Loves and kisses (used between male and female amateurs – less common now).

160 metres: The 1.8 MHz amateur band. Popular for short distance groundwave communication during the day and longer distances at night.

300 ohm: The characteristic impedance of ribbon cable at one time popular for TV antenna installations.

630 metres: The 472 kHz amateur band used by experimenters.

807: Popular tube (valve) used in homebrew medium-power transceivers from the 1930s.

2200 metres: The 136 kHz amateur band used by experimenters.

6146: Popular tube (valve) used in the last generation of transmitter power amplifiers before solid state took over. A pair will produce over 100 watts.

4 FINDING THINGS OUT

With only a sentence or two per definition, I can't hope to give more than perfunctory treatment to each. Except for the simplest you'll likely need further information to pursue a sub-interest that appeals.

How do you get it?

The first step is to read articles and watch videos by those who have done it. Note the equipment and skills needed. Learn the terminology and key words. Then search Google to unearth further information. Particularly read beginner experiences – some articles by those more experienced gloss over obstacles (e.g. computers not talking to hardware) that can appear insurmountable. A little preparation lets you ask better questions, avoid beginner mistakes and the purchase of unsuitable or unnecessary equipment.

Almost every sub-interest has a 'scene' that you may wish to associate with to learn more or ask questions. Devotees of special interests are normally welcoming of those wishing to join it, especially if they've made the effort to learn the basics. Involvement is particularly important for a facet like club contesting or microwaves where you are invariably arranging activities or expeditions with others.

If you're in a city of a reasonable size there's likely to be others that you can tag along with for your first experience of a particular sub interest. Some may be associated with local radio clubs, although the larger special interests have state or national clubs of their own.

If you're in a less populated area or have a demanding work schedule another way to engage with the masters is through online forums. The major sub interests have categories on forums attached to qrz.com or eham.net. Alternatively look for smaller online forums devoted just to that facet, Facebook groups or email lists. These can sometimes be hard to find or inactive. Check recent posts on websites and in other forum discussions for links to good ones.

Once you find and subscribe to likely groups read archives of past posts. Pay special attention to FAQs or 'sticky' posts. Use the search function if available – someone just last week might have asked and had answered the same question you wanted to pose.

There are right and wrong ways of asking questions online. A question asked well is a courtesy to those who may answer it while one asked poorly may get no or unsuitable replies. The knowledgeable people you want to hear from are neither mind readers nor have time to write book-length answers.

Specific questions that can be answered in a paragraph tend to get the best replies. The 'How long is a piece of string?' type don't. Sometimes it's useful to include steps you've taken to find the answer yourself or briefly mention options that you know people will mention but which are unsuitable (e.g. a huge antenna if you have little space). The manner of asking (e.g. tone and spelling) is also important, since you are seeking favours from strangers who are in no way bound to answer.

Examples of bad questions

* what is best arial for all bands amatuer radio??

* What is best HF transciever

* What time day can I work dx i call and call no one come back ever think they ignor me

Examples of good questions

* My yard is 20 metres long but narrow. I can fit a 10 metre tall mast in the middle. What is a good antenna for 40 and 80 metre contacts up to about 1000km?

* Seeking suggestions for an only home HF transceiver for a returning ham. Needs to be 100 watts and preferably new. $1500 budget but would prefer some left over for a portable QRP rig later in the year. The IC718 or IC7300 seem OK but the FT450 looks small for my fingers. I don't need FM, 6 metres or a band scope screen. What would you get?

* What are the best times to work Japan from midwest USA on 40 metres? Do any of you regularly achieve this with 100 watts to a ground plane vertical?

It's the nature of the internet that some answers will go off-topic, some will be from people with less experience than you while others will be gems. Follow up books, website links and videos suggested from the latter. And don't hesitate to ask for an elaboration if there's something about a reply you don't get.

Knowing terminology for each facet is essential. Much of it is in this book. Just enter words in Google or YouTube and you'll find more relevant pages.

Background research, finding where the knowledgeable people are and asking the right questions. These are the keys to learning about a new facet. Follow them and one day you may become the expert beginners seek to ask.

5 ABOUT THE AUTHOR

Peter (mis?)spent his youth at rubbish tips, taking apart given radios and TVs and building electronic projects that mostly did not work. He avoided soldering until figuring out that new solder on a shiny tip works better than reusing solder gathered from the chassis of vacuum tube radios.

Milestones included the construction of a crystal set, discovering shortwave broadcasting on a tube receiver and a simple 'electronic organ' from a Dick Smith Fun Way book. Hours were spent putting wires into springs on a Tandy 150-in-1 electronics set. Amazingly some wires could be pulled out and the project would still sort of work with only half the parts in circuit.

Two back to back AM/shortwave radios led to the discovery of amateur SSB activity and a novice licence. The following year was spent building transmitters no one heard. A one tube crystal-controlled CW transmitter from the 1973 ARRL Handbook provided the first contacts – mostly CW/SSB cross mode on the 3.579 MHz TV colour burst crystal frequency. The value of frequency agility was an early lesson and various VFOs were built, most of them drifty.

Further construction enabled more bands, more modes and smaller gear. Projects included a 7 MHz VXO CW direct conversion transceiver, 2m FM portable transceiver, and a 14 MHz CW transmitter for Cycle 22, then near its peak. Later favourites included HF DSB and SSB transceivers (often using ceramic resonators, ladder crystal filters, NE602s and BD139 transistors) and phasing SSB equipment.

Limited space led to experiments with magnetic loops and HF pedestrian mobile. The joys of the latter (along with the perils of a trailing counterpoise) were first discovered with a converted Johnson Viking CB on 28 MHz. This was mounted in a carpeted chipboard box with battery and 1.5 metre whip. A move to a beachside suburb brought further HF portable and pedestrian mobile activity which remains an interest to this day.

Peter is a prolific writer and video producer with items on the web and YouTube.

6 THANKS

Thank you for reading.

I invite you to share your comments and thoughts.

Facebook

Goodreads

Twitter

If you enjoyed this book and would like more please visit
http://www.vk3ye.com

There's many articles, projects and ideas on various facets of
amateur radio.

Also, should you wish to browse Amazon please click on the link
from my site above.

This referral helps support the site through a small commission on
any purchases made at no additional cost to you.

7 OTHER BOOKS BY VK3YE

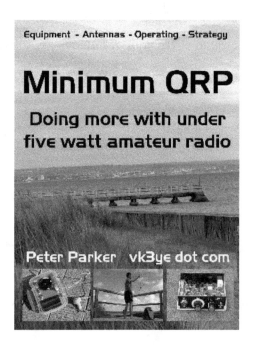

Minimum QRP: Doing more with under five watt amateur radio
contains tested strategies for low power success on the HF bands.
Equipment, antennas and operating are all covered in detail. Its
crafty tips for working the most with the least should become more
valuable as solar activity declines in the next few years.

Minimum QRP is for a broad worldwide audience. Newcomers, the more experienced and those returning to amateur radio are already benefiting from its contents with brisk early sales since its release.

Minimum QRP is available as a Kindle e-book. You can read it on a portable e-reader, your home PC or other device. Packed with over 200 pages of information it's yours for under $US 5 or equivalent. A paperback edition is also available in some countries.

Further details, including a table of contents, list of reviews and ordering information can be found at vk3ye.com or by searching the title on amazon.com

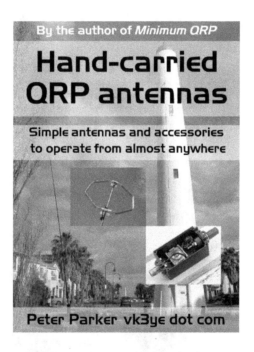

By the author of Minimum QRP

Hand-carried QRP antennas

Simple antennas and accessories to operate from almost anywhere

Peter Parker vk3ye dot com

Hand-carried QRP antennas is the book that takes the mystery out of portable antennas. After inviting you to assess your needs, it discusses the pros and cons of popular types. Its style is brisk and practical with almost no maths.

Many ideas for cheap but good materials suitable for portable antennas are given. Beginners and those returning to radio after a break should especially find this section handy.

Finally, there's construction details on a variety of simple but practical antennas and accessories suitable for portable operating. All have been built and tested by the author over almost 30 years of successful QRP activity.

Hand-carried QRP antennas is available as an ebook readable on most devices (free software available if you don't have a Kindle). A paperback edition is also available in some countries.

Further details, including a table of contents, list of reviews and ordering information can be found at vk3ye.com or by searching the title on amazon.com

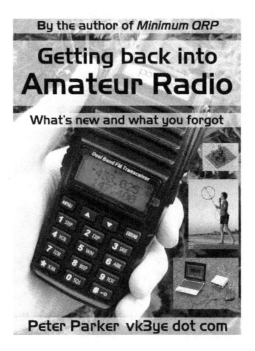

By the author of *Minimum QRP*

Getting back into
Amateur Radio

What's new and what you forgot

Peter Parker vk3ye dot com

Getting back into Amateur Radio is for those who have let their radio activities lapse but are considering a return.

In a clear, conversational style, it brings you up to date with what's changed and what you may have forgotten. In almost no time you'll be tuning the bands even if you don't yet own a receiver.

Getting on air is easier and cheaper than ever before, with a huge range of modes, bands and activities to choose from. You may even be able to become relicensed without sitting another test, depending on your country and documentation.

Ideas on antennas to use are given, with special help for those with limited space. And if you've forgotten operating procedure there's a refresher course on the various way to make contacts in this book.

Getting back into Amateur Radio is available as an ebook readable on most devices. A paperback edition is also available in some countries.

Further details, including a table of contents, list of reviews and ordering information can be found at vk3ye.com or by searching the title on amazon.com

What is now out there and how do you start?

Both questions are answered in *99 things you can do with Amateur Radio*.

It's an ideal primer for the beginner. It tells you things your class instructor probably didn't have time to cover. Try some of the facets suggested. Be amazed with what you can do even with an entry-level licence and simple equipment. Or, if you've been licensed for a while *99 things you can do with Amateur Radio* makes a good refresher on new modes and challenges now available.

99 things you can do with Amateur Radio is available as an ebook readable on most devices. A paperback edition is also available in some countries.

Further details, including a table of contents, list of reviews and ordering information can be found at vk3ye.com or by searching the title on amazon.com

For news on these and future books please subscribe to VK3YE Radio Books on Facebook or VK3YE's channel on YouTube

CPSIA information can be obtained
at www.ICGtesting.com
Printed in the USA
LVHW080007240722
724262LV00010B/358